D1542055

Television and the Moral Imaginary

Television and the Moral Imaginary

Society through the Small Screen

Tim Dant
Lancaster University, UK

First published 2012 by
PALGRAVE MACMILLAN

Palgrave Macmillan in the UK is an imprint of Macmillan Publishers Limited,
registered in England, company number 785998, of Houndmills, Basingstoke,
Hampshire RG21 6XS.

Palgrave Macmillan in the US is a division of St Martin's Press LLC,
175 Fifth Avenue, New York, NY 10010.

Palgrave Macmillan is the global academic imprint of the above companies
and has companies and representatives throughout the world.

Palgrave® and Macmillan® are registered trademarks in the United States,
the United Kingdom, Europe and other countries

ISBN: 978–0–230–23481–9

This book is printed on paper suitable for recycling and made from fully
managed and sustained forest sources. Logging, pulping and manufacturing
processes are expected to conform to the environmental regulations of the
country of origin.

A catalogue record for this book is available from the British Library.

A catalog record for this book is available from the Library of Congress.

10 9 8 7 6 5 4 3 2 1
21 20 19 18 17 16 15 14 13 12

Printed and bound in the United States of America

To the students who have talked about television with me, the teachers who have helped me talk about it (especially Jack, Gail and Stephen) and Mollie, who mostly watches it with me

Contents

1
Introduction – The Small Screen and Morality

I wonder whether Aristotle would have enjoyed watching as much television as many people in modern societies do.[1] He tells us that the arts give pleasure to people through mimesis: 'Imitation comes naturally to human beings from childhood (and in this they differ from other animals, i.e. in having a strong propensity to imitation and in learning their earliest lessons through imitation); so does the universal pleasure in imitations' (Aristotle 1996: 6). He was writing about poetry and drama, but television is today the medium that mimetically reproduces the life that humans directly experience as actuality and as fiction. The arts (including comedy, music and drama), the telling of history, the reporting of news and spectacles such as sports events are all mimetic forms that appear on television and give pleasure. Mimesis is not the same as a 'copying' or a 'mirroring' actual behaviour; it is always a representation. Even the most detailed and accurate audiovisual representation of live, actual events can never be direct duplicates of the events themselves, which, however faithfully captured, lose their smell, feel, depth and all-around-ness. Mimesis is a process in which something of 'reality' is always lost, and something is always added by the intervention of human action, and television's increasing capacity to appear to represent the real does not stop it from being mimetic.

Television – by which I mean simply 'viewing at a distance' – is *the* medium par excellence for mimesis, because it merges the capacities of all other media and incorporates all arts. Even, although very occasionally, poetry. As well as moving images, television gives out synchronised ambient sound, the visual channel can show still images and text in a variety of combinations, and the audio channel can be supplemented with music and commentary. A group of audiovisual media (including movie film, analogue video, digital video and interactive video 'games')

1

broadly share these mimetic qualities, and all can be carried on television devices in the home. In contrast, the printed word and image have a material substance that is relatively fixed and unmoving, although they clearly create a mimesis that operates over a distance and prefigures television. But the electronic media carry audiovisual imagery that is immaterial and has no substance beyond its electronic trace, which, once captured, can be replayed and repeated, and with digital media the repetition can be precise and apparently endless. Television is cheap, ubiquitous, easy to consume and accessible to all, and these characteristics, combined with its mimetic capacity, give it great social significance as a form of communication in late modern societies.

As such, I want to argue specifically that television has become the prime medium for sharing morality and dispersing the mores, the general ways of being and acting, throughout a culture. It is the mimetic force of television that is important because it is able to *show* rather than just tell what people do, and what the consequences of their actions are. This means that viewers watch matters of moral importance without thinking about it that way; what the viewer regards as 'entertaining', 'amusing' or 'interesting' is so because of its relevance to their life, showing something of what is important or showing a contrast to their ordinary lives. I want to argue in this book that the particularity of the form of television as a medium creates for its viewers a 'moral imaginary', a repository of ideas about the possible ways of living in the world. What these possibilities might lead to is included in the imaginary as are values about different actions, but they are absorbed more or less unconsciously; they are not like ethical rules or simple statements of 'should' or 'ought'. The moral imaginary is a way of sharing ideas across a society that can bring together groups of people, and it is a cultural space in which changes in mores and moral ideas can be rapidly passed on without instruction or formal education. Other media – for example, literature, newsprint, radio, film and the Internet – have not disappeared. They still exist independently as means of communication, and each has a role in contributing to the moral imaginary of modern societies, but they have all become more or less drawn into an alliance with the medium of television. Literature is dramatised for television, books of all sorts inspire television, newspapers review and reference television as do radio and the Internet, but they all also carry stories that are represented audiovisually on television. Films made for cinema have long afterlives as they are repeated on our small screens and sometimes are even shown first on television.

This book is not just about television as such but is about how the small screens in our homes have become a key way in which a modern society tells itself just what it is. The one-to-many communication that is organised into segments characterised by synchronised moving images and sound, and which is available on small screens in our homes, is a potent channel for socialising a society's members into its mores and morals. In this book, I want to explore what this might mean by using ideas from sociology to reflect on a range of issues, including some that are traditional in the literature on television. Television's characteristic mode of communication through the playing of moving images and coordinated sound requires neither an intended structure nor a systematic decoding by its viewers. Much of its content can be absorbed without conscious understanding and can be made sense of through watching its flow. Its various segments or programmes contain many values and ideas, but there is no obligation on the part of viewers to accept or adopt any of it. Indeed, the reasons why it is so important as a means of socialisation are that the flow of programmes changes over time, and the values and ideas they carry also change. The difference between segments – typically between different programmes – means that there is no unifying value system that underpins everything shown. This would be different where there is an institutional attempt to organise what appears on the small screen for a particular purpose – as happens, at least to some extent, with close state control over what is broadcast to ensure support for a political or religious order. But even tight institutional control over what is produced in the medium cannot control how it is made sense of, cannot prevent dissenting interpretation.

One important aspect of the interpretive fluidity of televisual content is in the phenomenological form of the dynamic audiovisual imagery that is characteristic of television and the small screen.[2] The content of the continuous present of television is watched as an alternative reality but one in which viewers use the same interpretative resources as they do to make sense of the paramount reality of everyday life. The continuous flow on the screen is complex, made up of some segments that are continuous in themselves (in its briefest moment, the 'shot') that are fitted together contiguously into larger segments (at its largest, the 'viewing session') to make the content of the small screen unlike any other medium. Flow is further complicated by the serial repetitions and references back and forward to other continuous sequences that occur at separate points in the flow (for example, the trailer that pre-views, the catch-up that post-views, the repeated programme and the replayed clip). Before setting out how this book will explain and expand on these ideas, let me explain first what I mean by 'morality' and 'television'.

Morality

What I mean by 'morality' is how people regard each other as a collective group – what they expect of others and what they think others expect of them. This is, by and large, not a rational or conscious process but one that for the individual members of society often seems to be 'natural' – a set of attitudes, values, ideas and feelings that they take for granted, as if it were built in and a part of who they are. Of course, sociology is precisely the attempt to articulate and make conscious in a rational, even systematic way, how being a member of a society is the product of historical and social process that make the constellation of society in any one space or time just what it is. While humans are part of nature, and the formation of social groups is natural, too, what distinguishes human beings from other animal species, at least as far as we can tell, is a reflexive sense of our collective being that shapes what we do; morality. Morality is, for most of us most of the time, an unknown known; we know what is right, but we are not quite sure how we know it. Things feel right or wrong, and our response to questions of how human beings should be and act often seems to be given within us, as if it were genetically inherited, a characteristic of the species, something beyond question. But then sometimes we do question the behaviour of others, passing comment and judging what other people do, and this questioning makes explicit what would otherwise remain implicit. The process of questioning, challenging and judging behaviour can be formalised and articulated as in a legal system, in the rules for behaviour in a sports or social club or the ethical systems for professions such as medicine and accountancy. Religious teaching and general education often set out rules for behaviour that, when they are taken up and followed, are effective as an authority for acting in a particular way. These are ways in which the members of a society have traditionally been socialised into the morality of their culture. In late modernity, although religion continues to be very important to many people, it is increasingly in competition with other sources of ideas about how to live. In pre-modern cultures, religion dominated the means of communication by which morality was passed on – writing, the schoolroom, the pulpit and the teachings of religious leaders. In late modern societies, the control over these means of sharing and distributing moral ideas has been secularised.

Rather than think of morality as a system of ideas that are coherent and connected, as happens in religions and ethical codes, I want to develop the idea of a 'moral imaginary' as something more fluid and inclusive. The imaginary will include images of possible ways for humans to act, to live and to be, ways that may be both right and wrong, depending on the context and the situation. Rather than being rules that are independent of context, the moral imaginary can hold variations on ways of being and include the different factors that might make what is wrong in one setting, right in another. Instead of absolute values, the moral imaginary includes the complexity of feelings, relationships and extenuating or ameliorating circumstances that allow actions to be judged in terms of the specific situation. These are not 'reasons', articulated in rational, causal, logical relations but dynamic images in which the complex flow of actions and events is grasped as a whole. At least in northern democratic countries, unlike a religion, television is not a coherent, systematic and articulate set of ideas and practices, and as a medium it is full of contradictory and competing messages about how to live. Since the medium is not in the control of a single organisation or institution, it is a conduit for many different perspectives and not just those shaped by a unitary system of ideas.

Television

Television, I want to persuade my readers, is a special medium precisely because it is consumed not as discourse, not as ideas and arguments that are systematically set out, cumulatively and deliberately. Some programmes may attempt to do this, but even these are characterised by what is distinctive about television as a medium; moving images with synchronised sound. Moving images are not in themselves arguments but are representations of life that both parallel the flow of life that we see unmediated in the world around us and the flow of images in our minds, as we reflect, fantasise, dream and imagine. When arguments are shown on television they are always accompanied by moving images, even if it is no more than a video recording of the person making the argument or articulating the ideas. But most television is not directly about arguments or ideas, it is about entertainment, pleasure, distraction, amusement, excitement, emotion and other forms of arousal, which means that it can never be reduced to a simple text, to the language of speech or writing. Watching sport, soap operas, thrillers, news, music programmes, drama, competitions, 'reality' shows and programmes in many other genres (and hybrids of all these genres) is not primarily about consuming ideas and arguments. Some documentaries, commentaries, thought pieces, political debates and discussion programmes – 'serious' television – are indeed about ideas and arguments. But one of the points I wish to make in this book is that the flow of television is consumed as a mélange of different types of programme all of which contain ideas about how we should live, and those that do not include clearly stated arguments are consumed alongside those that do.

Television is indeed more than just moving images because it is the juxtaposition of sound and images, and the putting together of segments of audiovisual images, that makes it 'television'. Most importantly, the sound and images that viewers consume are complementary channels of information that are combined for consumption by the senses of hearing and seeing, in the same way as we experience the world around us. When someone speaks on the small screen, we see the eyes and expression, the hand gestures and the bodily composure in the same way as we do when we are co-present with someone speaking. When an object moves on the small screen, a car, say, we hear the sounds of the engine and the tyres, and they are as indicative of the pace of the vehicle's movement as when we see a car move in front of us on the street. In addition to the ambient sound derived from what we are seeing, the aural channel can carry extra sound such as music that provides emotional resonance and a voice-over commentary that provides interpretation. On their own, they would have quite a different communicative effect, but these extra sound channels supplement the primary televisual form of moving image and ambient sound. Although virtual reality simulation promises something even closer to the sensory experience of the real world, the audiovisual communication of television is currently the most immediate form of communication available to human beings, short of co-presence. Its multi-sensory complexity and its photorealistic and audio authenticity give it the capacity to represent the world as closely to our human experience of the real world as is possible.[3] We could say that the medium of cinematic film was doing this some time before television and with greater photorealistic and audio authenticity, but television has three things that make it a more immediate medium. Firstly, it can represent reality as it happens – it can be *live,* and even when it is not, it carries with it, in a way that a cinema movie never can, the implication that it might be or could be live (as with a pre-recorded performance in front of a live audience or a show with a short transmission-time delay). Secondly, it is consumed within the *domestic* setting, emanating images as light in ordinarily lit surroundings in a way that integrates it into the flow of everyday life, a mundane, even habitual part of ordinary living. Movies are seen within the cinematic institution of a darkened theatre, with projected images, rows of seats, queues, payment and tickets for entrance – unless they are watched at home on a television. Thirdly, television is broadcast as a flow that is usually consumed as a *viewing session* rather than as a discrete unit as a feature film at the cinema is. The viewing session may involve a series of programmes, some only partly watched perhaps, with other sensory information and activities integrated and overlapping.

This book takes as one of its topics 'television'; the combination of Greek (*tele* 'far') and Latin (*visio* 'sight') words indicate a device that enables viewers to see things that are not currently present in front of them. But television is not what it was. The technological development of the equipment involved has changed what the material form of the television is taken to be. For many people over the age of thirty, the word 'television' means a piece of furniture in front of a semi-circle of comfortable seats in the main living room of homes in the industrial-ised world. The television I first watched as a child was a twelve-inch model with three control knobs (volume, contrast and brilliance) and veneered wooden doors – I remember watching reports from Dallas on it the day President Kennedy was shot. The electronic equipment it contained showed a dynamic black-and-white visual image on a small screen that had curved corners and a bellied front, with synchro-nised sound coming from a speaker underneath. Although nowadays much bigger, even as large as sixty-five inches diagonally, the televi-sion screen has always been 'small' in comparison to the 'big' screen of the cinema. The addition of recording equipment – initially video recorders but now DVD players and DVRs (digital video recorders) – increased the range of electronic devices that play sound and moving images in the home as television. Early domestic games machines and computers were designed to borrow the television screen as a monitor, but nowadays television may borrow the computer or games console screen for showing programmes. The most important change, however, is undoubtedly the arrival of digital technology, and with it the ability of computers to be used as televisions and of the Internet to carry tele-vision channels and individual programmes. 'Television' is no longer a dedicated piece of furniture but is now a medium that uses a variety of platforms, including desktop, laptop and tablet computers and even smart phones. The quality, definition and colour accuracy of small, flat and rectangular screens are very different from the television 'goggle box' of the 1950s, and there has been a proliferation of screens within the home. Some televisions have 'picture in view' on the main screen or remote controls with their own screen for previewing a second channel, and users can look to a second screen on a mobile device (laptop, tablet, smartphone) to simultaneously gather information or communicate with other people. The content on the second screen may be directly related to what is showing on the first screen – a blog or Twitter feed that makes 'live' even a recorded broadcast. The viewing experience may no longer be stereotypically with the family gathered around a set in a living room as household members have their own screens in different rooms, but watching alone can be made sociable by telephone or onscreen textual communication. What is surprising, given the increased consumption of screen-based media that are not televisions, is that consumption of broadcast television has been sustained and even increased. Sales of television sets in the UK increased from 4.75 million in 2002 to 9.55 million in 2010 – it appears that the purchase of sets and the increased viewing of them are linked to the introduction of technical advances that have improved viewing quality.[4] Ofcom research suggests that 'Online content complements TV viewing' rather than having to compete with the Internet and other screens in the home (Ofcom, 2011a, p. 38 see also *Economist*, 2010).

Themes

Before I introduce the contents of this book, let me tell you what you won't find in it. Firstly, I have decided not to use illustrations to accompany the programmes I will mention. This is because they would always be still images of limited size in black and white; a very different medium to the one I am writing about. Jeremy Butler's (2010) recent book on television style makes excellent use of series of small images to explain many of the stylistic features he is describing, and other writers on television have also made good use of images. But I decided that my sociological interest is in the way the whole medium works, and so I have referred only very briefly to specific television programmes, trusting that many readers will know of these or similar programmes. If they do not, they will find Internet sites with images, synopses, clips and even whole programmes that will give a much better flavour of the televisual experience than I can in a book. I am conscious that this book operates in one medium, published writing, to comment on another medium, television. Secondly, for those who are keen on 'television studies', this book will be frustrating because I will not explore key areas such as the institutional context and the political economy of the media that have come to be recognised as crucial for 'TV Studies 3.0', not only to understanding how the media work but also what its political significance is (Miller, 2010). I want here to consider television primarily as a medium of communication that has sociological importance rather than critique its institutions, regulations and financing even though these can affect how it fits into society. Thirdly, small screens show material that I will not try to comment on here: text and hypertext, still images and webpages and closed circuit television of all types. Fourthly, I will not attempt to discuss audiovisual material directed to a specific purpose such as advertisements, pornography, music videos, 'how-to' videos and video games. Much of the analytical argument I wish to make could be applied to these important communicators of social values, but they deserve more focused and direct attention than could be covered in a book of this size and type. Fifthly, surprisingly for an author who has published books titled *Knowledge, Ideology and Discourse* (1991) and *Critical Social Theory* (2003), this book neither treats the content of television as ideology nor mounts a critique of its content. There is an established critical theory approach to television (Adorno and Horkheimer 1979, Adorno 1991 [1954]; Kellner 1981, 1990, 1995) that has much to commend it but is also limited in its approach (see Nelson 1986). Although I do discuss some ideas from critical theory in Chapter 6, in this book I wanted to engage in analysis of the medium of television from a different, predominantly sociological and phenomenological perspective that I have used more recently to open up those aspects of the experience of material culture that are overlooked by critical theory (Dant 1999; 2005). One future project is to tackle the difficult relationship between critical theory and phenomenology as tools for understanding culture. But in this book I have simply avoided the issue.

This book is arranged to be progressive and cumulative with familiar and basic ideas clarified first, while the more abstract and unfamiliar ideas are to be found in later chapters. Those who want to find what I have to say about particular issues, such as the public sphere and where my concept of imaginary comes from, may want to use the contents page, the next paragraph and the index to move about the book less sequentially. Here I will mention some of the writers from sociology, philosophy and cultural studies whose ideas I have engaged with so that readers interested in particular theorists can easily find where they are discussed. But throughout the book, my aim has been to bring these often abstract and difficult ideas into conjunction with the sorts of programmes that one can easily see on any small screen.

In Chapter 2, I will explore how some very basic philosophical ideas of morality – virtue, duty and fairness – can be used to make sense of television programmes. Using a layman's gloss on the ideas of Aristotle, Immanuel Kant and J. S. Mill, I will show the relevance of ideas about morality in making sense of the content of television. I will turn to socio-logical ideas of morality in Chapter 3 and will question what role tele-vision might have in contributing to the moral order of late modernity through promoting a 'mediated solidarity', contributing to collective effervescence, creating moral events, depicting mores at home and abroad, documenting diversity and closely observing interaction that is morality-in-use. Here some traditional sociological ideas about morality from Emile Durkheim, Graham Sumner and Morris Ginsberg will lead on to more recent ideas of morality from sociologists including Thomas Luckmann and Zygmunt Bauman. Having established some important sociological features of morality, I will turn to television as a medium in Chapters 4 and 5, beginning by exploring its difference from cinema, developing Raymond Williams' concept of 'flow' and considering how technical advances have enabled a distinctive 'televisual' style in the ways discussed by John Thornton Caldwell and Jeremy Butler. In Chapter 5, I will develop a phenomenological understanding of how television is perceived by viewers and will draw on ideas from John Ellis as well as Henri Lefebvre, Edmund Husserl and Alfred Schutz to argue that we use the same interpretive resources for watching television as we do to make sense of our everyday world. In Chapter 6, I will recognise some of the key ways that we can understand television impacting on social life: as a mode of direct interaction using John Thompson's concept of 'mediated quasi-interaction', as a part of everyday life with reference to Henri Lefebvre, Paddy Scannell and Roger Silverstone, as an extension of the public sphere as originally discussed by Jürgen Habermas and as a means of socialisation in the light of social theory from Eli Chinoy to Peter Berger and Thomas Luckmann. In Chapter 7, I will return to the issue of how television mediates morality by considering media ethics and moral panics and the representation of distant suffering as discussed by Luc Boltanski, Keith Tester and Lilie Chouliaraki. The chapter goes on to consider just how television represents reality, and although the critiques by Jean Baudrillard and Slavoj Žižek are particularly helpful in making sense of recent televisual events, I suggest that often television is best understood as a form of witness as explained by John Ellis, John Durham Peters and Paul Frosh. The difficult idea of an 'imaginary' is the focus of Chapter 8, which explores how the term has been used by Jean-Paul Sartre, Jacques Lacan, Cornelius Castoriadis, Charles Taylor and Benedict Anderson and argues that it is through the role of viewers as spectators that television contributes to the contemporary moral imaginary. Combining insights from Adam Smith's idea of the 'impartial spectator' and Jacques Rancière's 'emancipated spectator' helpfully expresses how I want to argue that television works as a moral imaginary. In my short concluding Chapter 9, I will comment on the risks of damaging or losing the valuable contribution that television makes to the moral imaginary of modern societies.

An evening in

To give some flavour of the complexity of moral issues and imagery that a viewer might encounter in the flow of a viewing session, I will begin by writing about some television programmes of the sort that one might watch in a single evening. Even though they are all on the same channel and are made to the same high standards and address a general (as opposed to a niche or specialist) audience, they are all very different. None is about morality, but each, as I want to show, raises moral issues in quite different ways and different contexts. From an arbitrarily chosen midweek evening (2 June 2008), my viewing session is a sequence from BBC One. In that month, the channel was attracting nearly 28 million viewers on average per day, and television viewers were spending 24 per cent of their average weekly viewing time of 24 hours, watching BBC One, a higher viewing share than any other single channel.[5] This was the first channel to broadcast in the United Kingdom and continues to provide a mainstream mixture of news, information and entertainment without commercial advertisements. Although funded through a licence paid by all those who own a television set in the UK, the British Broadcasting Corporation is autonomous in the management of its affairs, and its obligations to both the state and its audience are set out in the Royal Charter that authorises the BBC Trust and an Executive Board to oversee it (DCMS, 2006). The Charter specifies that the BBC 'exists to serve the public interest', and its public purposes include 'sustaining citizenship and civil society', 'promoting education and learning' and 'stimulating creativity and cultural excellence'. The BBC is in competition with commercial broadcasters, and while its mission is 'to inform, educate and entertain', it needs to justify its public role by ensuring that it has sufficient viewers. Like all broadcasters in the UK, the BBC is also subject to the broadcasting code, which covers things like bias in reporting, intrusion into privacy and breaching standards of public decency.[6]

I have chosen a sequence of programmes broadcast to coincide with the middle of an ordinary weekday evening, a traditional period for family viewing when the competing demands of work, school, eating and sleeping are at a minimum so leaving space for peaceful leisure. The five programmes – *The One Show, New Blues, EastEnders, Panorama* and *Crimewatch* were broadcast over a three-hour period between 7pm and 10pm, the core of prime viewing time that reaches its peak at 9pm. These five programmes had been prepared for broadcast on television and were 'original' in the sense that they were not written as novels, plays or stories that were then adapted – nor were they 'feature' films made for cinema showing. In broadcast media, a 'programme' refers to a discreet and contained broadcast event lasting a specific time during the day. Programmes typically have a beginning, middle and end (marked by theme music and titles against a graphically devised visual background and closing credits with music) and have identified makers (for example, presenters, directors, producers, editors). They are usually scheduled to begin on the hour or half hour and last for blocks of time, usually of an hour or half an hour, although sometimes short programmes are for five or ten minutes. Compared with the variable length of novels or feature films, television programmes occur in a very strict temporal sequence with very rigid time spans.

Each programme has its own pace, style, structure and genre and addresses specific topics that are more or less linked together. However, there is a continuity between the structure of the programmes and that of the schedule; each programme is broken into sections that, while they are contiguous with one another, are also connected across time through frequent references forwards and backwards. Each section is further fragmented into editorially constructed segments and even within segments there is further fragmentation into scenes and then into individual shots. Programmes often extend beyond the boundaries of their opening titles and closing credits. Announcements, listings and trailers warn viewers of what is coming later in the evening, later in the week or on a broadcaster's other channels. This means that the regular television viewer is seldom surprised – just about every programme they watch has been previewed in a way that sets their anticipation. Between and sometimes within programmes are mini-programmes lasting from a few seconds to a few minutes. The classic form is that of the commercial advertisement, and the mini-programme is part of the contiguous flow of unlike elements in the viewing session that often contributes to the serial connection between programmes at different points in the flow. When the mini-programme is a 'trailer', it does not have a narrative structure so much as show dramatic events and key characters. Programmes often begin with a 'taster' sequence of what is to come later, much like a trailer, or with a 'reprise' of what happened in earlier episodes of a serial – these sequences come before the titles. Through trailers, tasters and reprises as well as repeats and clips, programmes spill out of the scheduled timings as some of their contents appear elsewhere in the flow. There were no commercial advertisements in the sequence I am going to discuss, but there were channel brand identifiers, or 'idents', over which announcers often give scheduling information. Although they usually include a logo (such as the letters identifying the channel in a distinctive colour, typeface or graphic form) as well as music and moving images, they may have a scene and unfolding action. In 2008, BBC One was using circling objects (kites, hands, breaking waves) in its idents to graphically represent the capital 'O' of 'One'.

The One Show

The One Show is a popular (4.05 million viewers in 2008) half-hour magazine show beginning at seven in the evening, with a regular team of presenters who introduce short, filmed and carefully-edited 'packages' with internal commentary and interviews. The presenters also interview guests and chat live in the studio, reflecting on packages and linking items. This episode included sections on: the continuing effects of a contamination of the public water supply in a small town, a look back at the career of a comedian with a new programme to promote, a report on the renovation of a network of nineteenth-century tunnels in Liverpool, a feature on spring and fox cubs and an interview with a man who bought a zoo and wrote a book about it. The programme items have a strong sense of temporality: now as against then (the water contamination, the comedian), the historical past (Liverpool tunnels), the time of the year (fox cubs) and the future (a new book, a new programme). The first two items were illustrated with previously broadcast footage, and a number of the items used interviews with individuals who were either witnesses to events or had an expert opinion to contribute. Some of these interviews were studio-based with the two presenters who maintain a genial and personal 'banter' and help viewers connect with past events and localised places through their own personal recollections. Rather like the feature pages of a newspaper or a regional magazine, the programme plays on a past that is in the living memory of many viewers and the geographical breadth of 'their' society. The 'now and then' structure of some items and the 'this time of year' temporal link of others counts as a particular form of news reporting; not urgent, important and emphatic news but reflective, items with a sentimental aspect. Imagery is important to stimulate recollections of people and events as well as provoke curiosity (the tunnels in Liverpool) and evoke sentiment (the close-up shots of fox cubs).

The presenters provide a friendly and informal base setting live in the studio; the ambience is light-hearted, interested, pleasant and slightly sentimental. Through their informal talk, they act as intermediaries between the silent and absent viewer and the tightly managed filmed packages, interacting with guests, making comments and raising the sorts of questions that viewers may have – such as 'What happened next?' The programme creates an interest in the lives and experiences of places and other people who are part of the same society but are almost certainly beyond the direct experience of viewers. The mores are those of a shared public space of interests that viewers can engage with: corporate responsibility (contamination of water), history (tunnels in Liverpool), nature (spring and fox cubs, the zoo), and biography (the comedian, the zoo keeper).

New Blues

Before the next programme, there are announcements and adverts for other programmes on BBC One and other BBC channels. The last links the upcoming mid-evening show, *Crimewatch,* to the programme that is just about to follow – *New Blues* – through a common concern with the law. *New Blues* is a documentary[7] in a short series about the recruitment and seven week-training of Police Community Support Officers (PCSOs).[8] The film follows three PCSOs undergoing training as case studies – they speak directly to the camera about their past, their expectations of the role and their experiences of training and induction. We also see their trainers talking to the camera and working with trainees. As with the filmed packages in *The One Show,* there is a narrative commentary that explains what is going on and links the various filmed sequences, but here there is no reflective studio discussion. A report questioning the value of PCSOs is mentioned as making them topical, but this programme is descriptive rather than analytical. It uses music with a strong beat, sharply edited camera shots from different angles and old-fashioned typewriter headings across the bottom of the image to structure and give the programme its particular look.

The film emphasises the standards of personal presentation, communication and fitness required by the PCSO service. The recruits learn self-defence, body searches and how they are expected to act; we see a change not only in the appearance of the recruits but also in their personalities and sense of self. The contrast between the expectations of the service and the values, dress and behaviour of the recruits before their training, mean that we observe their moral career as they transform into PCSOs. The values of close shaving, the right shoes and the physical capacity for self-defence within legal limits are not necessarily those that viewers are expected to adopt; but to maintain law and order, which includes moral order, the recruits apparently have to demonstrate these standards.

As the programme is ending, but before the credits have finished rolling, they are shrunk into a box on the screen while a new commentator tells viewers what is coming next and what is available on other BBC channels. Immediately after an 'ident' for BBC One, a new mini-programme begins – a news update with a woman presenter, standing with papers in her hand. Behind her, television film is used to illustrate the news items, many of which have a remarkable continuity with topics briefly mentioned in the *New Blues* programme – reports of people plotting to blow up a plane, a teenager stabbed and advice for parents on teenage drinking. Other news items include the security of savings in a building society (a topic relevant to the later *Panorama* programme), news of a local man telling police when he found a gun in his son's bedroom, and money being raised for a local cultural event. The regional weather report rounds off this short news section that ties the broader issues raised in programmes to local and timely ones.

EastEnders

If these two early evening programmes are different in style and structure, they have in common the reporting of events using moving images and spoken words to enforce their actuality.[9] They are followed by *EastEnders*, a long-running soap opera first broadcast in 1985 which is clearly fictional with actors playing characters in contemporary costume on realistic film sets and delivering lines learnt from a script. Soap operas have a series of overlapping story lines that draw in the cast of characters in various groupings. The common location – in this case, the fictitious Albert Square in 'Walford', London – is where the characters live and many of them work; their paths and the storylines of their lives criss-cross in space and time on the screen. In June 2008, the edition of *EastEnders* broadcast on Mondays had the highest viewer rating of any BBC One programme with an audience of 8.7 million – the only higher-rated programme broadcast that week was *Coronation Street* (ITV1), also broadcast on a Monday with over 9 million viewers.[10]

In this episode of *EastEnders,* there is a series of interconnected stories, and each has a moral dimension: no one helps the postman when the wheel comes off his cart (they all pass by), a young beautician is in conflict with her employers and her customers (she ends up taking money from the till and walking out... as does one of the customers), a family who has won the lottery plans how to spend their money (when the father finds he has lost the ticket, they throw his things out), a mother bolsters her son who feels inadequate, a couple have adulterous sex during their lunch hour (putting his back out interrupts them), a young man is locked out of his house wrapped in just a towel (women laugh at him and take advantage of his gaucheness), a child steals a flowerpot and gets caught (her mother makes her return it and gives her a new one) and a young woman who flashes her money while buying cocaine later gets mugged. Theft, sex, family disharmony, violence and drugs are all tackled in the half hour. The show is narrative driven, but the moving images allow for the characters' anger, delight, frustration, lust and depression to be shown as well as the consequences of their actions. *EastEnders* is followed by a mini-programme showing themes and characters for upcoming late-night BBC programmes aimed at young people.

Panorama

The swirling kites of a BBC ident are changed for the circling globe of the *Panorama* ident; first broadcast in 1953, it is the longest-running current affairs programme and has set a high standard of investigative reporting. We see the anchor, Jeremy Vine, introducing himself, the programme and the story – apparently live in real time – asking whether Britain's property market is 'finally grinding to a halt'. A series of snatches from the programme make a mini-programme giving us a flavour of the upcoming filmed and edited report with the theme of 'us' and 'financial worries'. The film begins with the reporter, Richard Bilton, driving around in a white van with the *Panorama* logo on the side. The van provides a metaphor of moving house as well the journey through the housing market and links the various components of the programme: interviews with house buyers and sellers including a series of 'cases' to highlight different situations, mortgage experts and estate agents; statistics and graphs; music to emphasise affect; to-camera presentations and facts; views of houses and locales.

A specially commissioned poll by *Panorama* enables the reporter to repeat a phrase used earlier by one of the experts; even in the aftermath of the economic crash of 2008, apparently 37 per cent would still be prepared to stretch themselves financially if it were for their 'dream home'. The plot – and the moral – of the programme become the 'unrealistic' desire for a 'dream home' that had been fuelled by cheap credit in the decade up until the financial crisis. Over a harmonica playing a slow, sad tune, the reporter says, 'They'd still go for it, even though we all now know the risks.' Those seeking to buy a house are encouraged by the sales culture to imagine their dream home, but this can lead to a mortgage they cannot repay. There is a moral theme of conflicting interests between those who are keen to own their own house and those who need housing but can neither afford to buy nor find appropriate rented homes. An expert panel with representatives from a local authority, the judiciary, a mortgage company, academia and the Citizens Advice Bureau offers pragmatic advice. They point out to one woman that she would avoid repossession and be 'financially miles better off if you rented and just gave up the dream of having... having your own home' The presenter sums up with a final moral message as he climbs into the van and puts on his seatbelt: 'With prices having their first sustained fall in ten years, maybe what we've learnt on our journey is that we need a change [slams door shut] of mindset, to stop seeing our properties as hot investments [clicks seatbelt] and to start seeing them [starts engine] as our homes [releases handbrake and pulls away].'

Crimewatch

Starting at the primetime of 9pm on my 'evening in', *Crimewatch* is a more complex programme that has been running since 1984 and ostensibly enrols the audience in finding and apprehending criminals. It is a popular show (with an audience rating of 4.83 million in June 2008) with a distinctive format that moves among live presentation in the studio, filmed interviews, dramatic reconstructions and clips from programmes broadcast in the past. Before *Crimewatch* can start, there are two trailers (one for a website linked to a different programme), and then the programme begins with the anchor's voice saying, 'Tonight, on *Crimewatch*' over a filmed reconstruction of a crime that is to come later in the show. It begins in the dark and the movement of people is unclear until we hear a woman pleading, 'Oh no, no don't, oh please don't, no', as she is being dragged away by two masked men. We cut to a well-groomed and well-lit woman speaking three-quarter face to an interviewer over whose shoulder the camera is watching. She says calmly, 'I thought that was the end, I thought, "That's the end of my life"', shaking her head from side to side. As we begin to move to the title sequence, we hear the anchor ask, 'Will you know the gang?'

The opening scene is intercut with militaristic, urgent drum rolls as it develops towards the title sequence that has a staccato trumpeted tune overlaying images and sounds of police cars and sirens, a helicopter, a large dog barking, an exploding car, swabs being taken of a door handle, police uniforms and vans and a young man running in tracksuit trousers and a balaclava. Graphic images, including fingerprints, rulers and a double helix, resolve into the two lowercase words: 'crime' and 'watch' coming together in a box. The anchor, Kirsty Young, moves through the studio in which groups of people are sitting at tables with papers and phones, her script in her hand. She sets the scene: 'Hello and welcome to *Crimewatch*. We're live, and the studio, as you can see, is packed, with detectives from around the UK, all hoping that your call will help solve a major crime.' The studio is separated into zones by see-through panels with white handwriting and stuck-on pictures (a staple crime drama prop). The detectives get on with their paper work while Kirsty summarises the main cases on the show this evening; the robbing of a businesswoman, the murder of a sheep farmer, the mugshots of 'most wanted' criminals, the taxi driver strangled and burnt in his taxi and the letter-bombing blackmailer who was caught. She moves around the studio, introducing her two co-presenters and gesturing to the relevant teams.

The programme presents particular actual crimes and the experience of their victims with a consistently serious and respectful tone, but within this broad context, the dramatic reconstructions of crimes owes much to the substantial volume of 'crime drama' shown on contemporary television. The actors are not shown full face, and action is sometimes filmed from a distance, through the dark or from strange angles and with dramatic music to emphasise the uncertainty and fear potential in a situation or an unfolding scene. Televisual cues from fictional crime drama (tickertape textual information on time and place, lighting, camera angles, music, script) are interwoven with the victims' accounts that lend the reconstructions a powerful veracity. There are four main sections, each looking at a particular crime and each lasting about ten minutes, and each is begun and finished live from the studio. Viewers are presented with a great deal of information in the sixteen sections of the programme over one hour, that cover a total of twenty-one crimes – each has some still photographic material, and thirteen have video material as well. This episode of the programme is looking for, probably, twenty-seven criminals. Detectives are filmed talking to the camera and are also interviewed live in the studio; they and the presenters show sympathy for the plight of victims as the rationale for their keenness to bring the perpetrators to justice. The presenters make regular pleas direct to the camera, asking the audience for help in identifying criminals and coming forward with evidence that might help the police. Textual information is displayed during the programme – the names of victims, relatives, police officers, phone numbers to call, the web address of the programme – to reinforce what viewers hear from the presenters. Much of the material is made available online (appeals, CCTV footage, reconstructions, wanted faces), and the programme makes much of the multi-media nature of the interaction with the audience through telephone calls, texts and emails.

Some segments use the poor-quality video from security cameras of incidents of crime – it is 'actual' rather than 'reconstructed'. Other segments use still images of wanted criminals, and there are frequent 'taster' clips previewing upcoming cases throughout the show to give the viewers a glimpse of the reconstructed drama of the cases to come. In the last third of the show is a sequence of three 'cases you helped to close' that reprises reconstructions and interviews that were shown in earlier programmes along with stills of the criminals who have been caught or convicted. As well as the reconstructions, video footage is used to illustrate the story wherever possible. In one case, video of the singer Jimmy Ruffin is inserted into a reconstruction sequence relating to a sexually-motivated murder that happened in 1971, in another, an amateur video of a murdered man dancing at a local gathering is shown with his face highlighted and news footage of a police press conference occurs in yet another. Still photographs, particularly police crime scene images but also family photos, newspaper photos and shots of headlines, are used where possible to illustrate the story. The imagery, moving and still, is used to reinforce the reality of the events being reported; they bring home the tragic consequences of the crimes. We are shown a clip of a murdered woman's husband who 'even did a special interview for *Crimewatch* as part of the appeal' but later turned out to be her murderer. Live broadcast television is linked to previously broadcast material and to online television where longer clips are available.

The morality of the criminals is never in question – they are all legally and morally reprehensible, although the presentation of the crimes does highlight circumstances (setting a taxi driver alight; robbing a frightened woman for very little money; an apparently motiveless murder of a man much loved and respected in his community) that lead to the outraged moral tone of the presenters and sympathy for the hurt of the victims. However, the moral import of the programme is really in the burden put on viewers to inform the police if they can identify, or provide any useful information about, any of the criminals. It is implied that failing to do so is failing in one's moral obligation to society in general and the victims in particular. In an interview in relation to a murdered Irish farmer, the family priest succinctly summarises the onus on viewers implicit in the programme as a whole:

> If members of the community know who did this, then I think they are duty bound to speak to the police. Someone's conscience must be bothering them, and they should have the courage to come forward and tell what they know. (*Crimewatch*, BBC One, 2 June 2008)

The anchor reinforces this message at the end of the programme: 'If you can help but you haven't called in yet, please do it now. It could be *all* down to you. I'll see you at 10.35 after the news. For now, bye-bye'. The feeling of equal and shared responsibility to do the right thing coheres the audience at the same time as they are being spoken to in a personal and interactive way. The vast majority of viewers are safe in the knowledge that they know nothing about these crimes and these criminals other than what they have just been told on television, but they are all included in the moral outrage, sympathy and concern.

The gloss on these five programmes is derived from a more detailed analysis; the moral significance of television is deeply embedded in the content of the audiovisual imagery that is ultimately impossible to render as text. Watching is the only way to get at it. In later chapters, I will refer to the content of a number of programmes – both individual episodes and series – but will abstract them from their context of the flow of ordinary television. What I was keen to do here was point to the moral issues raised in different programmes (a fine-grained analysis would have been more detailed and specific) and to describe the segmented nature of programmes within the segmented flow of a viewing session.

2
Morality on Television

From morality to the small screen

Television creates what I will call a 'moral imaginary'; the various aspects of other peoples' lives seen through the small screen swirl in an imaginary realm that is shared among people in a variety of ways. I will develop this idea of a moral imaginary later, especially in Chapter 8, but here I want to show simply that television programmes can have moral content that fits with the traditional themes of morality addressed by philosophers. The philosophic tradition has tended to discuss morality in terms of the intrinsic qualities of individuals (virtue), the principles by which their conduct is guided (duty) or the consequences of individuals' actions (seeking greatest good for the greatest number), and the content of television programmes can be seen as fitting with these approaches. This approach has been used as a way to make sense of literature (see, for example, the pieces collected in Pojman, 2000) and, more recently, computer games (see Schulzke, 2010) but not, surprisingly, television. The three themes I want to discuss – the good, the dutiful and the fair – would be, in principle, discoverable in any programme, but some are more relevant than others. What follows is not intended to be sophisticated philosophical debate but simply to draw out the connections between the narrative logic of television programmes and some well-established philosophical ideas about morality – virtue, duty and liberal utilitarianism. However, before I show how these themes are the stuff of television programmes, it is important to recognise that the issue of moral impact of television has been a contentious issue since if first began.

The immoral effects of television?

The theme of this book – television and morality – is not as original as I first thought. From its beginning, there has been concern about the immorality of the content of television, but the moral effect of the medium of television has also been discussed a number of times in the last thirty years. Social psychologists, especially those interested in developmental psychology, recognised that children would respond to the morality of television programmes. Bandura's famous 'Bobo' doll laboratory experiments are still cited as evidence of the moral effects of audiovisual moving images (Bandura, 1968, 1978 – for an overview see Gunter, 1994, p. 170) even though all they show is that children can learn new ways to play with dolls from watching films. The personal and psychological effects of television, especially the showing of sex and violence, have been thoroughly researched through field experiments, correlational surveys, longitudinal panel studies, experiments and intervention studies, but no consistent or persuasive demonstration of moral effects has emerged (see reviews of the literature in Cumberbatch et al., 1987; Felson, 1994; Gunter, 1994; Gauntlett, 2004). Despite the limitations of laboratory experiments for understanding what are increasingly recognised, even by psychologists, as sociocultural processes, social cognition theorists have continued to try to isolate the 'effects' of television (Berkowitz, 1964; Bandura, 1994; Krahé et al., 2001; Ferguson, 2011). However, social psychology has shown that viewers, including children, bring their own moral values to making sense of narratives (Zillman and Bryant, 1975) and that a sense of what is just and right affects their enjoyment of television (Raney and Bryant, 2002).

Within communication studies, George Gerbner and his colleagues (Gerbner, 1970; Gerbner and Gross, 1976) have argued that television has an impact on public attitudes and the value system of a common culture, not through a simple cause-and-effect model but by what they call 'cultivation'. How violence appears on television, they say, is inflected – or stereotyped – for the class, ethnicity, sex and age of both perpetrators and victims. What is shown does not necessarily tell us much about what people think or do but through analysing the key indicators in mass media messages, research 'will tell us much about the shared representations of life, the issues, the prevailing points of view that capture the public attention, occupy people's time, and animate their imagination' (Gerbner, 1970, p. 81). The cultivation theorists argue that television viewers develop a sense of heightened risk and uncertainty in the common culture, which leads to their dependence on established authority and acquiescence to state power. Viewers' fear of violence – the so-called 'mean world' effect of feeling that one is living in a violent society, that other people cannot be trusted and they are just looking out for themselves – varies according to the education and reading levels of viewers (Gerbner et al., 1994, p. 30). But viewers' impression of violence is not groundless; as Raymond Williams pointed out, each of the societies in which the traditional research on television violence was conducted '...was at the time engaged in violent action – some of it of exceptional scale and intensity – which had been authorised by the norms of the society' (1974, p. 117). The idea that the attitudes and values of the public were significantly affected in a malign way by television was widespread in the late 20th century, and the ex-advertising executive, Jerry Mander, argued that television should be eliminated precisely because it is such an effective medium of persuasion: 'I learned that it is possible to speak through media directly into people's heads and then, like some otherworldly magician, leave images inside that can cause people to do what they might otherwise never have thought to do' (Mander, 1978, p. 13). The damage of television to society was seen as personal, affecting how people thought and limiting their reflective and critical faculties as against a culture based on reading (Mander, 1978; Postman, 1985).

By the mid-1980s, media researchers and sociologists recognised that television as a medium might be having what Anthony Smith (1985, p. 2) calls 'big effects' and exerting social influence, but it was unclear just what those effects were. Establishing anything like a causal link was very difficult, but this did not stop an emerging public debate about the appropriateness of depictions of sex and violence on television (see Chapter 7) or continuing strong claims of effects by social scientists. Some of the effects were not negative (Smith, 1985, p. 5, quotes Meyrowitz, 1985 as an example), and some commentators began to suggest that television might contribute to society. Newcomb and Hirsch (1985), for example, floated the idea of television as a 'cultural forum', and Parsonian sociologist Victor Lidz specifically argued that '...television shows participate deeply in American *moral* culture' (1985, p. 27).[1] For Lidz, the secular moral order of America was constituted by a series of beliefs about the constitution, law, economics, freedom and education as well as morals relating to the family, the person and the community. The content of television programmes, he argued, reinforced a middle-class, conventional morality that, though secular, revealed its foundations in Puritan notions of self and conscience through the depiction of heroes whose reliability, trustworthiness and 'dependable devotion to collective ends' demonstrated their 'disciplined character' (Lidz, 1985, pp. 276–7). The idea was that television did not determine cultural values but provided a resource for viewers as their values took shape: 'Viewing serves as a generative symbolic practice, a conventional, socially marked occasion devoted to the production, by viewers, of peculiarly "cultural" apprehensions' (Saenz, 1992).

The position that Lidz and Saenz develop on the role of television as a means through which a moral order is generated is close to the argument of this book; what we viewers see on television does not have a direct effect or even necessarily exert an influence, but it does offer a range of 'narratives, consumer choices, moral predispositions and selected rituals of conduct' (Saenz, 1992, p. 43) that we may adopt, adapt, criticise or reject as components in our 'implicit knowledge'. Kenneth Gergen (2002) writes of the 'moral project' that is pursued through technological interaction in late modern societies, and researchers in different countries have described how this project works. In his study of Finnish viewers, Pertii Alasuutari (1996) shows how following the characters in stories can be a moral education, but at the same time there is a Finnish 'ethical realism' deriving from its Protestantism that resists programmes that are overly romantic or suggest that everyday life should be easy and so puts actuality programmes further up a moral hierarchy. From their content analysis of the moral messages, statements and subjects in the narratives of television programmes in the Netherlands, Krijnen and Meijer (2005) argue that, far from warranting a moral panic, they are: '...a useful tool in...raising and educating morally mature individuals who can act as responsible and socially involved citizens' (2005, p. 371). Gay Hawkins (2001) has argued that new programme formats have been part of an 'ethical turn' in Australian television and offer an ethics of everyday practices around cooking and gardening and an ethos of critical responsiveness through thoughtful documentaries. These are interesting ideas about television and morality, but they derive from a cause-and-effect model of social processes that is at odds with my concept of the moral imaginary. The most important sociological work on morality and television has been around the issue of distant suffering, and I will explore it in more depth in Chapter 7 (Boltanski, 1993; Chouliaraki, 2006, 2011; Ignatieff, 1985, 1998; Tester, 1999, 2001). For now, let me explore how the narrative content of programmes provides a contemporary resource for exploring moral issues, in much the same way as theatre and literature have for several millennia.

Virtue and the good

Aristotle's ethical system was oriented towards happiness, a general good not linked to any single aspect of living, but a complete and self sufficient end that we choose for itself. Happiness is a state arrived at through living: 'human function is activity of the soul in accord with reason or requiring reason' (1999, p. 9). It is not a momentary state but a cumulative one and is achieved through living and acting in accord with virtue. Virtue is a capacity of character rather than of action, and it has to be demonstrated by being chosen to guide actions – it is choiceworthy. This means that what appears to be a good action may not be virtuous if it was embarked upon by chance or following some irrational impulse. The demonstration of virtue depends on knowing the right action and then following it through for the right reasons. Put in this way, it is easy to see why, despite Aristotle's belief that animals have souls, he did not think they could be virtuous or achieve happiness. The donkey or the dog cannot be virtuous because they lack the reflective capacity that could lead to the formulation of actions intended to achieve particular ends. Human actions to be properly recognised as demonstrating virtue must be voluntary, an act of will and chosen for reasons.

The idea of virtue as an aspect of character that cannot be completely or adequately described in the abstract, lends itself to stories of heroic deeds. To undertake a series of actions to achieve an end that is recognised as good, is the stuff of stories from Greek myths to contemporary movies to children's adventures. Television has provided an excellent vehicle for the demonstration of heroic action because it can summarise action and outcome through narrative structure while heavily dramatising the actions and its effects. Whether it is the fantastical character of *Doctor Who* on children's television or the brave and adventurous members of the *Spooks* team, heroes act in the interests of others, often taking risks to realise the happiness of others. We are shown the operation of reflective thought – the reasons for action are often set out clearly in the script – and we see the happiness achieved by the beneficiaries and, as importantly, the happiness of the hero in having acted with virtue.

The hero who exemplifies virtue and strives for good is often a loner but sometimes has a sidekick and may even be the leader of a small team. He (it is usually a he) may have super powers (as does, for example, The Doctor in *Doctor Who*, or *Superwoman*) that parallel those of myths and legends and dramatically transcend those of the mundane run of human beings. Those characters who are 'good' are often starkly in contrast with those who are 'bad' and whose actions may be unequivocally evil rather than mere human failing. As a child growing up in England, I watched both the early episodes of *Doctor Who* (I even remember hiding behind the sofa...) and the more unequivocally exciting and heart warming *The Lone Ranger*, a series of television films made and originally broadcast in the United States in the 1950s with an eponymous hero whose persona was designed to demonstrate virtue. He helped those who were victims and not brave – a woman, perhaps a widow, a storekeeper or farmer with no capacity for violence – but was being bullied or exploited by a group of men, usually with a cunning or evil leader. The Lone Ranger would chase off the bullies, rescue the trapped child or recover the lost property through a series of actions that were calculated to achieve the desired end with the minimum of physical hurt to anyone. The base characters, the 'baddies' of the cowboy genre, were to be handed over to officers of the law but were frequently subject to public shame and humiliation in the process.

The Lone Ranger had a number of props that made him instantly recognisable but also acted as symbols of his virtue; a beautiful white stallion called Silver, who pranced dramatically in the opening sequence, a tightly fitting 'smart' grey suit (in those days, television was black and white), silver guns in black holsters, a white hat and a black mask. The 'bad' characters wore black hats and slouched with sneers on their faces and a crumpled demeanour. As the Museum of Broadcasting Communication website puts it:

> The Lone Ranger exemplified upstanding character and righteous purpose. He engaged in plenty of action, but his silver bullets were symbols of "justice by law," and were never used to kill. For the children's audience, he represented clean living and noble effort in the cause of fighting crime. His values and style, including his polished manners and speech, were intended to provide a positive role model. (Smith, no date)

He was virtue incarnate. He sought no reward other than to right wrongs and went on his way when the power of good was restored. For Aristotle, seeking to be seen to be good would be to demonstrate a shortcoming in virtue because the wish leading to the action would be for something other than the good outcome. The man behind the Lone Ranger's mask had no identity and could not gain from being seen to be virtuous. He used silver bullets to remind himself to use them sparingly. With the benefit of cultural distance, we might carp that the Lone Ranger exploited his 'faithful' friend Tonto, a native American who never took a lead in choosing a path or shaping the action. Neither the horse nor Tonto was able to demonstrate virtue; like the obedient Silver, the 'Indian' was simply a support for the main character. Most other characters demonstrated little virtue, and although a familiar plotline was the man who had slipped into a bad character and who was redeemed by being shown the error of his ways, the most wicked were often revealed to have double-crossed their accomplices.

The Lone Ranger demonstrates some of the virtues that Aristotle lists: courage, moderation, justice, generosity, mildness of temper, truthfulness, easy grace and proper judgement.[2] His actions in response to the moral transgressions of others were always measured and carefully planned. The lawlessness of the rural west of the developing United States of 130 years ago was the setting for *The Lone Ranger* – and many other cowboy films and television shows – in which the courageous and upstanding individual could make a difference by skilfully using force to achieve good ends. But the notion of the unequivocally good hero demonstrating virtue through simple deeds fitted with a post-war culture in which virtue could be associated with the reasoned and restrained use of force. Things have changed over the course of television history because heroes are no longer as clearly virtuous as they were, or at least appeared to be, in its early days. High drama, often fantastical drama, is the predominant mode that explores virtue and evil most explicitly, and Tim Kring's four-season series called simply *Heroes* (NBC in the US, 2006 – 2010; BBC Two and Three in the UK, 2007 – 2010) exemplified the genre. It began with twelve ordinary people who discovered they had superhuman powers and progressed as they came together in various combinations to fight evil – personified in an elusive character called Sylar – and to save collective humanity. These 'heroes' were flawed, making decisions that turned out to be wrong or were based on emotion rather than reason. In reverse of *The Lone Ranger*, evil was personified in one character, but virtue was the more fluid category appearing some of the time in some of the many heroic characters.

The virtues of the hero are most clearly shown in fictional characters, but the demonstration of courage and skill similarly emerges as a theme in some actuality programming such as those focusing on sport or certain professions (for example, rescue workers, firemen, paramedics). These sorts of programmes show the personal qualities of body, mind and character that are honed and directed through training and ethics to achieve a good end. Sports programmes show the meaning of fairness in competition, humility in victory and being a good loser, but above all they show the virtue of being a winner. Reality game shows, such as *The Apprentice* and even *Weakest Link*, also show the virtue of winning, often at the expense of team members, who are turned into competitors to be done down as the game proceeds. Competitive sports also provide an opportunity in which the less-than-virtuous character can be exposed; for example, cheating, the professional foul, using performance enhancing drugs. In documentary shows about the paramedics and police, some of those they have to deal with are shown on camera acting in ways that clearly lack virtue.

A critique of virtue ethics is that virtue is relative to the culture in which it is recognised; the shared approval of a behaviour within a society is what defines a virtue. Alasdair MacIntyre points out that what are listed as virtues in the writings of Homer, Aristotle and the New Testament are not the same, and none of them includes the virtue of constancy to be found in Jane Austen's writing, or Benjamin Franklin's virtues of cleanliness, silence, industry and the desire to acquire (1981, pp. 181–5). One of the themes of debate in both religious and philosophical writings is the hierarchy of different virtues, yet they are always ranked according to the general values of a particular culture. MacIntyre's own solution to cultural variation is to locate virtues within what he calls 'practices', which are socially valued and shared activities: *'A virtue is an acquired human quality the possession of which tends to enable us to achieve those goods which are internal to practices and the lack of which effectively prevents us from achieving any such goods.'* (MacIntyre, 1981, p. 191 – emphasis in the original). It is then a matter for the social group to identify the goods it wishes to achieve and the practices by which they might be achieved. A whole sociological conundrum is contained in MacIntyre's 'us' – who are 'we', and how do 'we' express our collective interests or the 'goods' we desire? The Lone Ranger might be given to the odd pithy statement about the importance of truth in his judgement of other people's character, but he is no philosopher, certainly not a moral philosopher. What are to count as appropriate practices is beyond question within the show, and the 'goods' – justice, peace and lawfulness – are taken as axiomatic. On the other hand, a contemporary show like *Heroes* is far more equivocal about what counts as virtue, and the special gifts of the 'heroes' are something of a burden as they strive to identify good uses for them.

Virtue ethics as presented by Aristotle does imply a universality – what is virtuous in Greek society of the 4th century BC is characteristic of all societies at all times – but this is clearly a problem for philosophy.

What is remarkable is that the Aristotelian system of virtues, despite cultural variation, makes so much sense in the 21st century. If there are arguments about their universality, they do not necessarily lead to a need to claim any extreme relativism. Martha Nussbaum (1992) identifies three arguments against virtue ethics that turn on this point; firstly that while Aristotle may have identified the broad issues, his solutions do not take into account the variability of cultural experience that will lead to different responses; secondly that the cultural values or 'grounding experiences' that identify the issues can vary; and thirdly, that it is possible to imagine a life where those cultural values are irrelevant because the grounding experiences do not exist. But Nussbaum does not give up so easily on the idea of a consistent human experience on which morality can be based. She resists the relativist retreat to culture and argues that there are fundamental aspects of human experience on which a virtue ethics can be built: mortality, the body, pleasure and pain, cognitive ability, practical reason, early infant development, affiliation and humour (Nussbaum, 2009, pp. 365–6). All cultures have to address these issues that are common to all humans in one way or another. What Nussbaum is doing is to reformulate some of the key virtues that Aristotle has articulated, so that courage is expressed as the need to face human mortality, and friendship becomes affiliation or the fellowship of other human beings. The reformulation takes into account some of the cultural transformations that make contemporary societies so different from Aristotle's while confirming some essential aspects of what it is to be human.

Duty

A second theme of morality that contributes to the content of television is the deontological judgement of an action that follows an obligation or duty that should be recognised by the actor. It may be the energetic and risky playing out of virtuous behaviour that makes television 'action' drama and sport exciting to watch, but such programmes also often show the value of a sense of duty to a profession, a skill, a team, the law or other things. An obligation to act in particular ways that the character would not freely choose is built into roles in police and military dramas as well as into actuality programmes like *New Blues* (see Chapter 1). Duty is also a recurring theme in soap operas and those programmes where obligation to family or friends or to a particular relationship is an issue that shapes the choices that characters make. The significance of duty may often be most identifiable in its breach as subsidiary characters spell out the failings of those who have transgressed in some way. The theme of duty and obligation also occurs in documentaries and 'reality'[3] shows that exhort viewers to act in certain ways, such as to eat healthily, to exercise or to treat their children or pets properly. Here the duty is to follow a proper way of life as set down by experts who are deemed to know better than ordinary people. There are endless ways in which this can be played out as, for example, in the many series of the reality show *Wifeswap*, broadcast since 2003 on Channel 4 in the UK. (The format has also been exported to the US and many northern European countries.) The programme mixes the members of families by getting the wife and mother (it is occasionally a husband) from one family to swap with her equivalent in the other. The duties and obligations of the role of 'wife' within each family become explicit as its members complain that they are not fulfilled. The swapped 'wife' has the opportunity to comment on the extent to which her new family meet her expectations of how they should act, and sometimes this involves reflecting on the limitations of her own family. Lessons are apparently learnt in what can, at times, be a rather brutal and bruising personal experience – not only for wives but for other family members. Despite the connotations of sexual unfaithfulness of the title, the programmes carefully avoid the sexual and emotionally intimate content of the family's lives. Instead, the focus is on the role obligations of various family members in relation to each other that are often given piquancy by the contrast in lifestyle, wealth and social class between the two families.

A dramatic presentation of the obligations and duties within families is offered by *The Sopranos* (HBO), a critical and popular success with an amoral character, Tony Soprano, as the leader of a New Jersey mobster family and a domestic family; much of the narrative force lies in the compromises needed to meet the different duties required for these two different families. The show ran for 86 episodes over six seasons between 1999 and 2007 in the United States but has become a worldwide hit as both a broadcast series and as a DVD boxed set. The characters in the show, including its 'hero', Tony Soprano, do terrible, illegal and immoral things and yet they are redeemed through their acceptance of duties and obligations to the domestic family and to the Mafia family of confederates in crime (the two families overlap as cousins and uncles are relationships in both). The tension between these two sets of obligations leads to Tony suffering from panic attacks, black-outs and depression, such that he seeks treatment from a psychiatrist. His character has courage and a strong sense of justice, a form of 'honour

amongst thieves' in which there are rules and obligations for relationships amongst mobsters that may mirror, but are separate from, the laws
of the nation state in which they live. Justice has to be managed outside
the courts through mob rule – this leads to summary punishments
for transgressions and to reciprocal acts of violence between different
groups, or 'crews'.

It is perhaps mischievous to suggest that the moral order of *The
Sopranos* can best be understood in terms of Kantian deontological
ethics – there is no evidence that this is what the scriptwriters or producers intended. Kant argues that it is 'good will', rather than happiness
or utility, that is the end of reason and so the focus of ethics – good will
is not a means to any other outcome (such as moderation, self-control,
sober reflection) but is an end itself: '...it is good only by virtue of its
willing – that is, good in itself' (2002, p. 196). He recognises that there
is something strange about putting absolute worth in a mere will, but
he argues that it is *'manifested when we act out of duty rather than inclination; only such acts have moral worth'* (Kant, 2002, p. 197 – emphasis
in the original). When in season six Fat Dom kills Vito, he may take
pleasure in the killing, but he shares the act with Gerry, for whom it is
just another task, and they are both following the instructions of their
boss, Phil, who hated the fact that Vito was homosexual. Like soldiers
who cannot be tolerated questioning the commands of their superiors,
their moral duty is to follow the commands of their boss. On the other
hand, when Carlo kills Fat Dom for taking his role in Vito's killing, he is
exceeding his duty because he is not required to act in this way. When
Tony orders the blowing up of the wire room (a communications hub
for Phil's crew), he is acting in accord with his duty – a duty to maintain the balance of power amongst the various crews within the family.
If he does not, then his crew and all their dependents risk losing their
status and so their livelihood and way of life. Many of the actions of
members of the gangs are breaches of normal moral principles, and they
do not even follow personal inclination or interests. That these violent
acts demonstrate 'good will' is difficult to accept unless we recognise
the 'good' to exist in sustaining the way of life of the members of the
family. A similar style of reasoning would explain the actions of soldiers
invading a foreign country whose 'good will' would be to root out and
kill terrorists and so neutralise a threat to the way of life of the society
in which they and their families live. The military analogy helps to
make sense of *The Sopranos,* whose leader is their 'capo', or captain, and
whose lowest members are 'soldatos' or soldiers. According to a BBC
news report, the 'ten commandments', a behaviour code for members of

the Mafia discovered during an arrest, include 'Always being available for Cosa Nostra is a duty – even if your wife is giving birth'.[4]

Underlying actions such as the killings of Vito and Fat Dom and the blowing up of the wire room are what we might recognise in Kantian terms as maxims; 'one must not be seen to be homosexual', 'one should respond to loss with an equivalent response'. Vito attempted to 'explain' his homosexuality in terms of 'confusion' following the use of blood pressure medicine, but this didn't work. Phil ordered Vito's killing because of his homosexuality, but Tony's response of blowing up the wire room was to respond to the loss of Vito's potential financial contribution to his crew. The sustaining of sexual, financial and power status is a matter for substantial and decisive action. As Fred Feldman explains, maxims are not necessarily conscious thoughts that accompany actions, but they are the ways that they might be explained after the event: 'If someone were to ask you to explain what you are doing and to explain the policy upon which you are doing it, you might then realize that in fact you have been acting on a maxim' (Feldman, 2009, p. 237). Viewers watching *The Sopranos* have to construct such maxims from what they see because they are not maxims that would work in most ordinary lives. Once used to the dramatic conceit of the show, the viewer can suspend the moral order of their own life and enter the moral order of the diegesis.[5] Feldman points out that although maxims are linked to particular actions rather than types of actions, they are general in the sense that they explain what anyone would do under these circumstances. Kant's account of the metaphysics of morals includes the idea of 'universal law', and he writes: 'That pre-eminent good which we call "moral" consists therefore in nothing but *the idea of the law in itself*, which certainly *is present only in a rational being* – so far as that idea, and not an expected result, is the determining of the ground of the will' (Kant, 2002, pp. 202–3 – emphasis in the original). This is a complex passage, open to interpretation, but it seems to be saying that for an act to be moral it must arise from a person's understanding of their duty under the law. They must be rational in that they can make sense of the law and understand what sorts of actions it requires. The operation of 'will', the voluntary entering into an act, is then moral insofar as it is motivated in a desire to fulfil one's duty to the law rather than to bring about particular circumstances.

Just how voluntary the actions of the mobsters are is not always clear, but for Kant's ethics they should be steered by the 'principle of will' and not by fear of reprisal or a desire for further ends (such as monetary reward or promotion within the organisation). Their actions should be guided by a sense of duty about how they should act rather than to achieve particular ends. If we read the 'law' to be the law of the Mafia rather than the law of the land or of religion, then it is easy to see how some of the characters, some of the time, are acting morally. The translators of Kant's writing on morality recognise in their substantial introduction this problem of applying Kantian ethics and consider the argument that a Nazi killing innocent Jews could claim that he was doing his duty. Their rather weak response is that such a misguided sense of duty is not 'duty as Kant understands it', but they are able to offer no textual reference to draw this distinction (Hill and Zweig, 2002, p. 33). One of the viewing pleasures of *The Sopranos* is in seeing how the characters (and makers of the show) balance the tension between the mobster law and a rather different morality in the context of the domestic family in which a duty of care, especially towards children, is uppermost. Some of the latter issues are mundane and familiar – such as dealing with children and their problems in finding a way in the world – and might crop up in any soap opera or drama. But they are often given an edge by the leaking of issues from the crime organisation into everyday life. For example, Tony Soprano is keen to guide his children away from a life of crime but cannot resist using his gangster connections to help them by arranging for his son a job on a construction site that is connected with the mob and arranging for his daughter's unsuitable boyfriend to be killed.

At the centre of Kantian ethics is the 'categorical imperative' that identifies what actions must be done to fulfil one's moral duty under the law: 'Act only on that maxim by which you can at the same time will that it should become a universal law' (Kant, 2002, p. 222). To act according to a maxim, one must also 'will' that it is one that everyone else shall recognise as the law. Philosophers debate whether Kant sees this generalising or universalising of a maxim as a necessary or sufficient condition and whether it refers primarily to actions or to the nature of maxims (Feldman, 2009, p. 241). However, a simple interpretation that does not address (and certainly does not solve) such problems, sees the imperative as being very similar to the so-called golden rule familiar from millennia of ethical debate, of treating others as you would like to be treated and not treating them in ways one would not like to be treated oneself. Perhaps the crucial difference is in the language that does not orient the consequences of one's own actions directly back to oneself. A mobster may reckon that it is reasonable to 'whack' (*The Sopranos'* favourite euphemism for murdering others within the crime world) those who step out of line with the Mafia code, and accept that they run the risk of being 'whacked' in return. But Kant's categorical imperative would mean that if a maxim of reciprocal 'whacking' were generalised, every act of violence would be met with an equivalent act of violence until the warring factions had eliminated each other. What we see in *The Sopranos* – even in the episode when the wire room is blown up – are periodic attempts at reconciliation and bringing about peace. This requires forgoing a reciprocal act of violent revenge in the broader interests of the organisation; duty is not just to the crew but also to the mob.

Viewers can enjoy a thrill at the wickedness of *The Sopranos* but are able to view it as a viable alternate social world that has a morality that can sustain the lives within it; they do not need Kantian moral philosophy to make sense of it. However, viewers cannot but confront the moral dilemmas that arise after watching it for anything more than a few minutes. The moral order created within the diegesis raises questions and problems with which the audience can engage; their curiosity about what happens next is linked to their interest in seeing the logic of the moral order being played out.

Fairness

Contests between virtuous and bad characters, between wilfulness and duty, make for dramatic television, but the fundamental idea in utilitarianism of being fair in sharing out pleasure and pain, of maximising pleasure for the greatest number and minimising pain for as many as possible, does not generate much dramatic action. Drama tends to revolve around the actions and choices of individual characters rather than the consequences for people in general; to create a utilitarian drama would be rather dull, if only because the most important activity would be the invisible mental one of calculating the consequences of actions. Utilitarian philosophy is most pertinent in the genres of actuality, especially documentary, news and magazine programmes in which the 'greater good' is often spelt out through the way that actual worldly events are framed and reported. Such programmes take for granted that viewers share with the producers a desire to minimise pain and suffering while maximising pleasure amongst all. Such a presumption underpins programmes reporting, for example, distant suffering caused by war, flood, famine or earthquake or the viciousness of a political regime or abject poverty (see also Chapter 7).

Originally writing in 1863 John Stuart Mill (1987) argued that the problem with Kant's linking of moral action to duty was that it could lead to terrible harm by focusing on behaviour, more or less without regard to its consequences for other people; the categorical imperative was not a sufficient safeguard against the harm caused to others by dutiful actions. Soldiers and gangsters who will that their criteria for acting according to duty are universal, accept the reciprocal rights of others to act in the same way, which then leads to continued destruction until one side is victorious in dominating the other. For Mill, happiness in the individual could not be regardless of the happiness of others, so each individual has a responsibility to the right of liberty for all – a principle higher than duty. The utilitarian Greatest Happiness Principle '...holds that actions are right in proportion as they tend to promote happiness, wrong as they tend to produce the reverse of happiness. By happiness is intended pleasure, and the absence of pain; by unhappiness, pain, and the privation of pleasure' (Mill, 1987, p. 278). This was the basis of Bentham's utilitarianism, which took a systematic approach to measuring pleasures in terms of intensity, duration, certainty, propinquity, fecundity, purity and the number of people they extended to (Bentham, 1987, p. 87). He also listed the kinds of pleasures – those of the senses, wealth, skill, amity, good name, power, piety, benevolence, malevolence, memory, imagination, expectation, association with other pleasures or the pleasure of relief from pain (Bentham, 1987, p. 90). It is individual people who experience pleasures and pains, but Bentham was trying to set out practical advice that might direct the efforts of government in maximising the good for its people. Appropriate intervention might increase the common good, but both intervention and punishment should only be used when it would lead to an increase in the sum of pleasure rather than simply to establish order or bring about retribution.

Bentham did not distinguish the relative worth of different pleasures, and in response to its critics, J. S. Mill developed a distinctively liberal version of utilitarianism in which 'happiness' is a variable good that can be realised in many different ways. He begins by pointing out the common tendency to treat morality as a natural faculty, as a sort of 'moral instinct' such that every human knows what is right and wrong. But he shows that morality is a matter of reason rather than perception and, although he locates moral sensibility in the individual, he argues that it is collective beliefs that tell us what actions have a positive effect on happiness and are the basis for morality. Experts, including philosophers, may come up with better ways of acting to improve the greater good, but human beings know, from what they have learnt from their society, how they should act. Instead of calculating and weighing each act for its impact on the happiness of all, people can draw on the accumulated experience of the species: 'During all that time mankind have been learning by experience the tendencies of actions; on which experience all the prudence as well as the morality of life are dependent' (Mill, 1987, p. 295).

Mill would probably have been very tolerant of television programmes in general, regarding them as largely trivial and of little concern to ethical philosophy – they contribute slightly to general happiness and seldom cause much pain. But certain sorts of programmes, such as documentaries that as well as summarising facts take a particular perspective that suggests a way forward, demonstrate a concern for utility and liberty according to a higher order sensibility that he would have approved of. For example, *The End of the Line* is a one-off programme based on a book by a journalist, Charles Cover, about the way fish are exploited as a human food. Cover is one of the experts who appear in the programme as a 'talking head' speaking directly to the camera and other academics and experts give testimony, such as the tuna fin diver turned 'whistleblower', Roberto Mielgo Bregazzi, who says, just before an advertisement break: 'What's at stake here is, an infamous minority of people making millions and millions and millions, [sniffs] by decimating a specie [sic]. Is that right? Is that moral?'

The programme argues that the future of fish as food is seriously threatened by the way that fish are caught and farmed. The ecological critique of current fishing practices is based on the need to balance the immediate desires of consumers against the longer-term availability of choices for consumers. The film is clearly trying to educate its viewers about the changing utility of catching fish; the greater good of maintaining the possibility of having fish in the sea in the future, some of which may be eaten, depends on reducing the individual pleasures of consuming fish, especially by very damaging means such as bottom trawling or intense overfishing. The 'whole' and the 'greatest number' are not here the members of a single nation-state society but include most people on earth; all those who are living and even those yet to be born. The programme spells out the contradiction of wasteful fishing practices that lead to the destruction of the very resource – wild fish – that humans want to eat. But because it takes five kilos of anchovies to produce one kilo of salmon, even fish framing depends on an impossibly rising demand for wild fish to feed farmed fish. Although a powerful critique of consumerism, the programme was shown in the UK on Channel 4 in a late-night series called *True Stories*, sponsored by Honda under the slogan 'Everything matters' and surrounded by advertisements for a range of consumer products. A multi-national collaboration of companies made the programme using footage and experts from many parts of the world.

Utilitarianism gives us a way to understand both the narrative logic and the televisual devices that are used by a programme like *The End of the Line* Although it can also be seen as a polemic, a campaigning document, it makes a series of rational claims, supported by academic argument, that the greatest happiness of the greatest number of people is best served by introducing restraint in the ways that fish are gathered for human food. Mill distinguishes between sensual pleasures – which to some extent are shared with animals – and mental pleasures such as those to do with beauty and music, which we might presume animals do not share. He recognised that some do not have the time, access or opportunity for indulging in or sustaining intellectual tastes, but clearly for Mill anyone who had experienced both sorts would recognise that the pleasures of the 'higher faculties' produce a finer quality of happiness than the more direct, sensual pleasures of the body (Mill, 1987, p. 280).

This one television programme, with its single theme of the risk of overfishing, confronts its viewers with a complex moral dilemma cast in the utilitarian mode. *The End of the Line* describes a range of pleasures that are primarily bodily and derive from catching and eating fish; the pleasure of hunting and catching the fish, the pleasure of satisfying hunger, the pleasure in the distinctive taste of particular fish and the pleasure from money earned by selling fish. A loss felt at the level of the higher faculties would be the pleasure in the diversity of sea life, a recurrent theme of the imagery in the programme which used a series of visually aesthetic tropes of sealife, reminiscent of the BBC's internationally successful *Blue Planet* documentary series about the world's oceans, first transmitted in 2001. The play of light on the silver tones of repetitive patterns of scales on the fish are set against the complex and moving blues and whites of sea and sky. Shoals of fish form dynamic patterns as their coordinated flow is sometimes sudden, sometimes gentle, while music is carefully chosen to enhance the mood. These pleasures of television are also threatened if overfishing means that there are no longer exotic fish or vast shoals to film, so the aesthetic gives emotional force to the argument being presented.

Mill recognises the category of a 'moral right' in which a claim or obligation that may not be formally recognised in law becomes apparent as a matter of justice within the principle of utility (Mill, 1987, p. 313). The right to the way of life fishing cod in Newfoundland threatened by quotas, or the West African fishermen, whose stocks of fish are being rapidly depleted by foreign trawlers, were presented as moral rights in the film. Less morally clear were the subtle pleasures of eating particular types of fish – bluefin tuna, for example – that, despite their endangerment, appeared on the menus of restaurants in capital cities. For Mill, happiness is not a single or simple thing, but the principle of utility has a social significance that comes together in the collective interests of everyone:

> ... laws and social arrangements should place the happiness or (as, speaking practically, it may be called) the interest of every individual as nearly as possible in harmony with the interest of the whole; and, secondly, that education and opinion, which have so vast a power over human character, should so use that power as to establish in the mind of every individual an indissoluble association between his own happiness and the good of the whole ... (Mill, 1987, pp. 288–9)

This passage links happiness to 'interests' and sets out the role of education and opinion, and it is not difficult to see a television programme like *The End of the Line* that promotes restraint, as being in the interests of the whole social group. The media, and in particular television, are crucial in the late-modern era for educating and influencing opinions amongst the population and in keeping their attention on the significance of the greater good. Left to their own devices, people may concentrate on their own happiness or at least that of those immediately around them. It is only through mediation that people can learn of the consequences of their actions beyond their immediate situation and can weigh up their happiness against the interests and good of the whole of society. Those media can include the words of others received in a variety of ways – public speakers, sermons, newspapers and books – but the audiovisual imagery of telemedia provide a powerful stimulation to reflect on the general good. Mill proposes a world in which 'The good of others becomes to him a thing naturally and necessarily to be attended to, like any of the physical conditions of our existence.' (1987, p. 304) The television can make especially present the interests of others around the world. *The End of the Line* provides a reasoned and sustained argument, supported by facts and figures, graphs and expert testimony about the extent, dangers, threats and appropriate response to overfishing. It shows the tension amongst the interests of fishermen, consumers, politicians, scientists and campaigners but challenges complacency and fixed attitudes. The scientists have collated the experience of humankind in relation to the fish stocks – they have gathered the knowledge that can inform the prudent actions of others.

Unusually for a television programme, *The End of the Line* also offers a manifesto of how we – the viewers – should act individually towards collective ends, and the programme finishes with three forms of appropriate action: reducing consumption and campaigning for responsible fishing and also for protected marine areas. Here we have a fundamentally utilitarian argument; instead of maximising my individual pleasure by eating what I want now, I should restrain my desire to consume in the interests of the general happiness of the wider community.

As well as making sense of programmes that are trying to persuade viewers of the utility of a policy, Mill's view of happiness can help us understand the pleasure of watching television as a pastime. He argues that intense pleasure only lasts for a relatively short time and it could not be continuous and sustained. We should not look for 'a life of rapture; but moments of such, in an existence made up of few and transitory pains, many and various pleasures' (Mill, 1987, p. 284). Human beings, he argued, need a balance of sufficient excitement and tranquillity – with enough tranquillity, they can put up with not much pleasure, and with enough excitement, they can put up with a considerable amount of pain. Different personalities will seek a different balance, and it will vary over the life course so that tranquillity can enable recovery and preparation for excitement. Television watching creates a bodily passivity that – as with the consumption of literature, theatre, film, funfair rides and recreational drugs – is more or less tranquil but can be exciting at the same time. The imagination is provoked into a quasi-experience that has some of the simulated qualities of an actual experience, while the sensori-motor apparatus of the body is more or less relaxed at the same time. Watching a carefully made television programme in which characters are involved in a car chase is exciting, though not in quite the same way as actually being in the car as it is being chased. Television can be a source of pleasure for the watcher at the same time as its moral effect indicates possible sources of pleasure and pain for people in a more general way.

Conclusions

Those early commentators on television as culture, Horace Newcomb and Paul Hirsch (1984), Victor Lidz (1984), Michael Saenz (1992) and George Gerbner and his colleagues (1994), were on the right lines in seeing television as a medium that lends itself to dispersing exemplary stories of moral – and immoral – action across a whole society. Cultural forms have been used for moral purposes since cultures have existed; myths, plays, poems, stories and novels have all been vehicles for messages and debates about moral issues. Television has tried showing philosophers discussing ideas – in the 1970s, Bryan Magee's *Men of Ideas* series were hour-long interviews with such key thinkers as Isiah Berlin, Herbert Marcuse, A. J. Ayer, Noam Chomsky and Iris Murdoch. But as Derrida remarks in a television programme, '...it is not possible to discuss a text like *Being and Time*, on television' (Derrida and Stiegler, 2002, p. 112). Philosophy does not work well *on* television because it is an abstract, discursive, textual and argumentative form, but I have tried in this chapter to explore the ideas of philosophers of morality in relation to what television *shows* us as examples of the moral consequences of human behaviour. It is no surprise that films and television should continue the traditional role of culture in showing examples that explore morality, and I hope I have persuaded you that Aristotle's notion of virtue, Kant's of duty and the utilitarian ideas of fairness and the greatest good for the greatest number can be helpful for interpreting what is going on in all sorts of different television programmes.

It would be interesting to engage in a rather more complex analysis than I have attempted here to find out whether virtue is demonstrated in a show like *The End of the Line* that appears to address the greatest good, or whether the concept of maximising happiness for the greatest number can help make sense of the *Sopranos* (which would perhaps be possible if it is taken to refer only to those within the extended 'family'). If the difference between these perspectives on morality is one of philosophical principle, rather than relevance to a particular story of life, are they then relevant to any culture at any time? Philosophies of morality tend to operate with absolute concepts that are intended to be applied universally. As I have suggested, the difference between the way the Lone Ranger acted and the way the twelve main characters in the first series of *Heroes* acted, indicates that the ways that 'heroes' are depicted on television has changed over the course of fifty years. No doubt, a similar shift in cultural representation of moral positions has occurred in other media (novels, plays, movies), but it raises the issue of cultural relativism. The meaning of any particular cultural product – whether on the large or small screen, whether performed or written – will vary according to the social setting in which it is being consumed.

How *The Lone Ranger* is viewed depends on the time and the place; even in the UK, where there never was a 'wild west', the show looks crude and corny today. (You can find extracts and even whole shows online.) No doubt, *The Sopranos*, originally broadcast in American English (with a smattering of Italian words and phrases), has been watched all over the world (subtitles in French, German, Polish, Greek, Spanish, Portuguese and Turkish are easily available online). The tolerance within a society of a second culture such as that of the Mafia as depicted in *The Sopranos*, with its own mores, values and correctional system, will vary according to the traditional values of nation, region, class, age and even gender of those watching. But if the way morality is presented on television is always relative to the social context of the viewer, does that mean that attitudes to torturing political prisoners, or female circumcision in one nation state must be tolerated by another on the grounds that they have a different moral culture? The issue of moral relativism, which seems to mean that one set of moral standards might be appropriate for one society but not for another, undermines the philosophical ethical systems that want to clarify things once and for all.

Steven Lukes has recently analysed the arguments for and against moral relativism and concludes that while it is reasonable to recognise different moral cultures, this does not mean that there is no basis for the judgement that certain norms are wrong:

> ...one can take the Kantian line of asking whether a given practice can be justified to all those affected, or one can take the Aristotelian line of asking whether it drags those involved in it below the threshold of one or more of the central human capabilities. Many ways of life – involving different forms of marriage and gender relations, for example – may pass these tests, but wife-battering certainly will not. (Lukes, 2008, p. 142)

Lukes is not going as far as Martha Nussbaum (whose work he discusses – see also my brief discussion earlier in this chapter) in arguing for a series of spheres of human life in which if sufficient capability is supported, fundamental experience will be similar and so principles of virtue can be derived. But he is arguing that while anything like a universal worldview is not sustainable, even within Nussbaum's 'capabilities' approach, there are grounds for recognising consistency in ethical responses across very different cultures at very different times. Lukes uses John Rawls' phrase of 'an overlapping consensus' amongst different cultures that allows for variation in morality between societies but still points to a measure of common ground (2008: 134). He suggests that literature from different cultures provides evidence of the existence of common moral ground, but cultural mediation through literature – and, of course, movies and television – can also promote shared moral views and attitudes, extending the moral imaginary across legal and linguistic boundaries.

Rather than being a traditional philosopher of ethics, Lukes is a social theorist with interests in sociology and politics, and he is cautious of dismissing the importance of relativist ideas. Sociology, well into the 20th century, tended to presume that the more complex and functionally refined social systems were somehow more resilient to threats of social collapse, but it has developed an epistemology that allows for a high degree of relativism, accepting that what works in one culture or society is not necessarily what is needed in all cultures or societies. The development of sociological approaches to understanding cultures and societies grapples with the variation in moral order amongst and within societies and offers a more nuanced response to moral relativism that I will develop in Chapter 3.

It is important that the viewer does not have to agree with the moral premises of the characters in a television show and may actually be alerted to the moral issues implicit precisely because he or she does not agree. What is more, the viewing pleasure of the unfolding narrative and sequence of images, of the action and the setting, do not in any way depend on agreeing with or accepting the values implicit in *The Lone Ranger* or *Heroes*, *The Sopranos* or *Wifeswap*. Most television does not present itself as a parable or allegory; it is not instructional and is open to interpretation in a variety of ways. *The End of the Line–* is unusual in being a programme that seems to demand that the viewer agree and act according to the values it presents – but even it has no power other than persuasion. Like all television, though, it raises issues and sets out different ways of living and of doing things.

3
Sociology and the Moral Order

From philosophy to sociology

Morality is about the obligations of the individual to the social group, and moralists over the ages have been keen to clarify what those individual obligations are. Philosophers and theologians have used the tools of systematic thought to consider how people should act and how we should judge the actions of others. I showed in Chapter 2 how some of their ideas can be used as a way to understand the content of television programmes, but one limitation in the philosophical tradition is that morality is treated as a matter primarily for the individual – if a person's way of thinking or being is right, the goodness of their conduct will follow. The philosophical tradition pays great attention to the cognitive aspect of morality; thinking precedes action and makes it what it is. Sociology turns the emphasis around so that morality refers to the behavioural standards of a society and derives from seeing what people normally do in that society rather than from thinking about what an individual might do.

'Ethics' can be a synonymous term for 'morality', but I would like to try to treat ethics as a system of rules that may be written down and treated as a code, and morality as the more general values relating to behaviour in a society. Ethics are often the rules of a particular social group, such as a religion or profession, and have a quasi-legal form but without the force of law, while morality is uncodified and allows for variation and interpretations. The emergence of ethics related to the conduct of business was of great interest to classical sociologists trying to understand the transition from traditional societies ordered by religion to modern societies in which business was more dominant (Durkheim, 1992; Weber, 1992). The concern with ethics at the end of the 20th century has developed particularly around the relationship between doctors and patients, and television has provided an opportunity to explore some of the key issues, such as medical intervention with very premature babies (for example, *Great Ormond Street*, BBC One, broadcast in three episodes in 2010) and the possibility of assisted dying (for example, *Terry Pratchett: Choosing to Die*, BBC Two, 2011).

The sociological tradition looks at morality as a feature of societies rather than individuals: the 'shoulds' and 'oughts' of moral awareness are invocations to act in certain ways that come from outside the individual, from society. As Emile Durkheim puts it:

> ... since we know that morals are the product of the society, that they permeate the individual from without and that in some respects they do violence to his physical nature and his natural temperament, we can understand ... that morals are what the society is and that they have force only so far as the society is organised. (Durkheim, 1992, p. 74)

Morals may shape a person's conduct, but when they do, it is because the requirements of living in a society have become internalised, and the link to actions is often through a learnt habit or way of acting. A child may respond to appeals to behave in a particular manner from their parents or other grown-ups, but soon the way to behave becomes internalised so that the teacher is able to say with conviction, 'You know you shouldn't hit other children!' This is also how morality is for adults in everyday life; they already know what they ought or should do in a given situation and often feel as if this is a personal quality that derives from some intrinsic capacity or even a moral gene. Similarly, the idea that someone might be 'evil' is often treated as intrinsic to the person; it is just how they are, as if they were made that way. Durkheim resists the idea of a 'moral nature' that is constituted in the individual and argues that the relationship between the individual and state that establishes rights in the person and a reciprocal moral obligation is variable, not only amongst different people but also as individualism becomes more pronounced, the state evolves and becomes more complex (1992, pp. 67–8). The sociological perspectives on morality of Emile Durkheim, Graham Sumner and Morris Ginsberg that I shall discuss in this chapter, share the view that morality varies from one society to another, across time and across spatial boundaries.

Being moral is a social relation between the person and the particular society he or she lives in. But the individual does have a degree of autonomy and choice; what one *should* do is not necessarily what one *wants* to do or is in fact *going* to do. The inclination of a person may be to satisfy their own wishes and desires, but morality constrains them to recognise that there is some higher order, beyond their own needs and impulses, that has different interests and requires a different mode of action. The culture of a society is more than morality, but morality is a cultural matter both as a topic for reflection and comment and in the way that the people in a society interact. For Durkheim (1992), 'society' was closely related to political society, the state and the geographical space of nations, but in a global mediatised world, 'society' has extended beyond these boundaries (Urry, 2000).

In this chapter, I will explore various aspects of the sociological idea of morality as a thoroughly social phenomenon not determined either by inner human nature or some extra-human power. I will begin by taking Durkheim as my guide with the overarching idea of a 'moral order', a moral system that is a feature of whole societies and constrains individual conduct, and then consider some features of it; from the solidarity of the social group and its energising through effervescence to the profane morality of everyday life and the ordinary interactions that make it up. These sociological aspects of morality progress from the macro to the micro moral order of society, and each can be seen to be linked to the content and style of television programmes.

Moral order

What is a moral order? It is the largely unwritten system of social mores and conventions through which the society is kept together as a coherent whole, and without it individuals would be pitted against each other and live in constant fear of what others might do. The moral order operates both as an internal constraint on the individual and as a publicly shareable account of the appropriateness of individual actions. The restraint on violence towards others, for example, based on a shared presumption of respect for the other person's motives and intentions, is an important feature of any moral order. Of course, there are special social contexts in which the moral order is differently articulated; a judo or boxing club, military training, war or the restraint of citizens who are in breach of the law are contexts in which violence is morally acceptable within particular ethical constraints. Rather than a single set of principles or 'master rules', for Durkheim the moral order lies in the norms, customs and mores of ordinary social life. These he sums up as 'ways of acting that we do not feel free to alter according to taste' (2002, p. 28) but which mean that each person does not have to engage in the 'incessant search for appropriate conduct' (2002, p. 37) – he or she simply has to do what others accept as normal. Durkheim often treats the moral order as equivalent to society, but he also recognises that there are variations within it, amongst different professions for example, and that the individual's relationship with it changes over the life course.

There is no moral value in behaviour that is directed towards personal ends; moral obligation is to the social group: 'Moral goals, then, are those the object of which is *society*. To act morally is to act in terms of the collective interest' (Durkheim, 2002, p. 59). The individual does, however, have an autonomous moral consciousness that reflects the 'duplex' character of human being as both a social subject and a free individual, both obliged by the social imperative and free to think and reflect on the reasons for it. It is in our interests to accept the constraints of the moral order and 'Only through our informed consent is it no longer a humiliation and a bondage' (Durkheim, 2002, p. 118). Moral consciousness is not learnt from reasoned reflection on principles but evolves through rational and critical engagement with the ordinary, situated, activity of social life going on around us. Society is made up of groups of individuals – family, community, nation, region, humanity as a whole – but has itself a status as a moral entity that evolves towards civilisation. The individuals contribute the 'moral power' and authority to society, which in turn provides its members with a 'storehouse of intellectual and moral riches' of which the individual on their own possesses only fragments (Durkheim, 1974, p. 54). To live within the moral order, people exert a measure of self-discipline, which is not about suppressing evil, combating sensual nature or taming some inherent badness, but is about maximising potential for a full life. For Durkheim, we can only achieve happiness and satisfaction by acting within the cultural limits of the moral order, and it shapes our hopes, feelings, appetites and inclinations, because excessive freedom would lead to chaos through the pressure to act differently or in new ways. Freedom for the individual only comes within the rules and limits we all recognise, because without them we would each fear oppression and exploitation by others. The flow of ordinary social life that we see on television – both fictional and actual – shows people living within the moral order, accepting its constraints without losing their individual identity. Much narrative content is about the ways in which the ordinary flow of social life is disturbed and shows transgressions of the moral order that attract sanctions, perhaps through a legal process or through the disapproval of other characters.

The legal philosopher Kurt Baier (1995) sees the moral order as 'anchored' to society and part of the culture that 'includes society's customs, traditions, manners, law, religion and morality' (Baier, 1995, p. 201). These are systems of sanctioned directives for members' behaviour and attitudes which are taught in the course of growing up and being socialised. Baier's sociological approach recognises that these different systems of valuing social actions are interconnected. Good manners are fundamental to the smooth running of social systems, and the different forms of interaction between people who are familiar, compared with those who are strangers, is a feature of many genres of television such as those described in Chapter 1. The process of socialisation into the mores of one's culture is often associated with education and religion but is increasingly undertaken through the media. For adults in modern societies who have left education, the media, especially the audiovisual media, continue socialisation into the rapidly evolving mores. Where religion may have a role, it is often deeply conservative and, by comparison, is limited in its moral outlook.

For Durkheim, the moral order was put under threat with the shift from a traditional society characterised by close relationships with people who shared the same life experiences, to a modern society characterised by the division of labour in which people undertook different roles but still needed to feel mutual obligation. What melded the traditional society was 'mechanical solidarity' expressed through the shared feelings realised in the collective or common conscience, 'the totality of beliefs and sentiments common to average citizens of the same society' (1933, p. 79). The processes of modernisation (industrialisation, urban migration, increased social density, the decline of religion, the increased importance of money) threatened the solidarity that was manifest in the shared feelings of shock and outrage at transgressions of the moral order. Durkheim suggested that a new mode of 'organic solidarity' was developing in which relations amongst people who had different lives and experiences followed rules of respect and obligation around transactions, especially those of employment, property and consumption. Since Durkheim was writing, mediatisation has extended the ways in which solidarity and the acceptance of a shared moral order has kept societies together – a sort of 'mediatised solidarity' has emerged in which values are shared through the media which communicates enough common ground to connect the members of societies, not necessarily as a single, coherent group but as various subgroups, with some values shared across subgroups.

To put it another way, the *conscience collective* finds expression through the media as well as through the outrage felt about real transgressions. Television news is less likely than tabloid newspapers to report crimes in a way that will provoke the *conscience collective*, but television drama often creates characters that inspire feelings of hatred and disgust shared by viewers who are both horrified and fascinated by what they do. Such characters are a staple of crime drama; they have few redeeming traits and are depicted for viewers to be shocked and appalled by their violent acts. An episode of the series *Single-Handed* (ITV, 2011) began with people chasing a teenager who was presumed to have robbed and killed an old man who was found dead near his cottage. Those chasing were outraged at the action of the teenager, who had absconded from a young offenders' training centre, and many other characters were also convinced that he was responsible for the death. In a later episode, the local police sergeant discovered that it was a different, middle-aged man who was responsible for the killing, and he had been taking revenge on the old man for abusing him many years earlier when he had been a boy in a local children's institution. The show plays with the collective outrage that viewers share with characters when they see the old man on the ground, and the teenager running away; it is a reasonable presumption that the teenager must be brought to justice. However, as is often the way with an unfolding narrative, the 'twist' of introducing another character with a justifiable motive for killing shifts the outrage to the old man, who had been a child abuser. His killer must face the law but should be treated with sympathy for the harm he has suffered.

Soap operas, too, will introduce characters whose excessive and wilfully immoral acts trigger an emotional response in the audience that expresses a shared conscience. For example, 'Dirty Den' in the long-running BBC One soap opera *EastEnders,* was unfaithful to his wife, fathered a number of illegitimate children and was a violent gangster whom audiences so loved to hate that he was 'resurrected' after he had been killed off once, just for him to be killed off again a few years later. The cold-hearted and cunning oil baron J. R. Ewing was a similarly celebrated character whom the audience of the American show *Dallas* (1978–1991) also loved to hate as he cheated and lied his way to wealth, adultery and corrupt power. Violence is a frequently recurring theme of television programmes; from children's cartoons in which anthropomorphised animals wield weapons with mortally devastating (but usually short-lasting!) effect, to the police procedural in which one murder is closely followed by another to draw out the cognitive work of detection. Our curiosity and concern about acts of violence make it a recurrent theme of documentaries, news reports and other actuality programmes as well as providing a narrative force in soap operas, hospital dramas and even comedy shows. This ubiquity of violence on television is because the use of violence, and its effects on victims and everyone else, are of great social interest; how it is dealt with is an indication of the moral order of the society.

If the *conscience collective* can be felt through the shared outrage at the representation of transgressions, both fictional and actual, the social relations of organic solidarity are also a common theme of television programmes. Moral order is not only about violence, and the sociologist Robert Wuthnow (1987) argues that the moral order of modern societies is shaped by the market:

> We associate the market with some of our most cherished values or moral objects. We also perceive the market in the context of symbolic boundaries defining realms of intentionality and inevitability…. An erosion of the market system may threaten the very fabric of moral commitment and challenge some cultural sources from which self-worth is derived. (Wuthnow, 1987, p. 79)

More than just a price mechanism, Wuthnow argues that markets enable freedom and choice and public participation in society as a whole. It is in buying, selling, working and consuming that individuals recognise and follow moral principles and obligations to other people beyond their immediate family and community. These are the relations of organic solidarity, which produce the cohering moral order of modern societies and are again shown in many programmes. How to exercise choice in buying music, clothes, cars and domestic technology has been a popular theme for television series that, in the UK, often emphasise lifestyle as much as engagement directly in the market. In *What Not to Wear* (UK – BBC One/Two, 2001–2007; US – TLC, 2003–2008), two presenters, having criticised the dress style of a woman (sometimes a man) through focusing on body shape and self-image, then gave them £2,000 to shop for new clothes following 'rules' set by the presenters. *Top Gear* (BBC Two, 1977–211, now with versions in Australia, Russia and the US), tells its viewers about new models of cars and does comparison tests while promoting a particular version of car consumption that emphasises the excitement of speed, power and manoeuvrability. Since its reformatting in 2002, the show has become famous for its middle-aged male presenters and their hammy, boyish humour and delight in machinery and its destruction. The *Gadget Show* (UK Channel 5, 2010–11, also Australia) reviews consumer technological devices – cameras, MP3 players, games machines, computers, home audio and more. The presenters in all of these shows judge not only the consumer objects – cars, clothes and gadgets – but also the consumers, showing approval for those who accord with the ethic of the show (fashion and making the 'best' of their body, speed and power, or exotic devices with new capabilities). In Wuthnow's (1987, p. 88) moral order, the value of freedom is expressed in the voluntary nature of action in the market, and the consumers who enjoy these sorts of programmes are free to watch or not, to follow the advice or not. Nonetheless, the programmes provide a standard of what is a 'good' product and a 'good' consumer, a set of symbolic boundaries against which the viewer can imagine themselves and their car or clothes being judged.

A different type of programme reinforces the moral value of the good entrepreneur or salesperson, the most famous of which is *The Apprentice*. While the formats of the consumer-oriented shows just mentioned originated in the UK and were then sold or franchised around the world, *The Apprentice* originated in the US with Donald Trump as the principal judge of entrepreneurial behaviour (NBC, 2004 –present) – the UK version has featured Alan Sugar in the key role (BBC One, 2005 – present). The show is a competition amongst volunteer contestants who have competed for the chance to appear and who are then tested and judged on their business skills; salesmanship, negotiation, leadership, teamwork and organisation. The contestants have to work in teams under a leader, and yet their loyalty to each other is in tension with their desire to win against the others. The demonstration of competitive spirit, determination and ambition are often the key to recognition, but still Alan Sugar as the judge of contestants' performance takes pleasure in criticising them and telling one each week, 'You're fired!', a catch phrase that emphasises the gladiatorial style of the show.

Mediated solidarity

Before the electronic media, mores and moral ideas were '... repeated from mouth to mouth, transmitted by education, and fixed even in writing. Such is the origin and nature of legal and moral rules, popular aphorisms and proverbs, articles of faith wherein religious or political groups condense their beliefs, standards of taste established by literary schools etc.' (Durkheim, 1938, p. 7). In electronically-mediatised societies, their distribution is wider across space and time and reaches more people within and beyond the nation-state society that Durkheim is thinking of. The moral order is the cohering force, the tacit agreement to live in accord that keeps individuals together in a society. Television programmes like the ones mentioned do not determine the moral order in any simple way, but they present moral issues and the symbolic boundaries of values and judgements about behaviour linked to violence and death, consumption and entrepreneurialism (as well as many other moral issues). The moral order is a 'normative system' involving 'reciprocal moral beliefs' agreed and shared by the members of a society to help to 'control impulses' so making social life more reliable and predictable because of 'common definition of a situation' (Gouldner and Peterson, 1962, p. 47). But it is not a fixed or stable system of values and television programmes such as those mentioned put forward values that are far from universally shared within the society; not everyone agrees with Trinny and Susannah's judgement about clothes or Jeremy Clarkson and his co-presenters' views on the simple pleasure of manoeuvring a fast car. The business acumen of the 'apprentices' is not so important as their performance in what are often, to viewers, rather trivial tasks. Nonetheless, entailed in each format are sets of values that are treated as 'normal' and not subject to question. Most importantly, these values are seldom articulated or stated; they are shown, in action and interaction, being taken up and acted upon. Presenters and key characters take them for granted, often reinforcing the sense of what is right in a fine judgement with which the viewer may not at first agree but is encouraged to consider. At least to some extent, to engage with the programme, viewers must acquiesce in accepting the moral values that underlie these judgements – those who cannot are unlikely to watch or become engaged in the programme's story. The mediated solidarity that television brings is not a new form, rather it extends both mechanical and organic solidarity across greater social and geographic distances. Mediated solidarity supplements the continuing interactional relationships in both the traditional space of the community and the modern space of the impersonal city. The television is often a focus for other media (newspapers, web sites and social media as well as face-to-face interaction) in which events and issues are reviewed and commented on.

Effervescence

In trying to understand the origins of the collective conscience in religion, Durkheim (2001) was impressed with the way that coming together could produce 'collective effervescence', a heightened emotional state shared by people who were acting together, traditionally in ritual situations and sacred events. 'It is the *collective efferves-cence* stimulated by assembled social groups that harnesses people's passions to the symbolic order of society. ... Collective effervescence, then, has the potential to substitute the world immediately available to our perceptions for another, more moral world' (Shilling and Mellor, 1998, p. 196). These events bind people together through a set of shared ideals and values and generate an energy that replenishes the social and moral order. In the traditional societies that Durkheim described, it was religious ceremonies that brought people together to refresh solidarity and moral cohesiveness. He pointed out that festivals, 'even ones that are purely secular in origin', can also have the effect of 'bringing individuals together, setting the masses in motion, and so inducing the state of effervescence, sometimes even delirium, that is not unrelated to the religious state' (Durkheim, 2001, p. 285). In late modern society, public events such as sports competitions, political rallies, music events, dances and festivals of all sorts bring people together and are occasions for heightened emotion. These occasions often include vocal and bodily gestures that are purely expressive, 'exuberant movements' that are 'performed strictly for the pleasure of performing and delighting in something like games' (Durkheim, 2001, p. 283). Religious events such as the Hajj and the Kumbh Mela, and political events such as gatherings in Tiannmen Square in Beijing and Cairo's Tahrir Square, are also opportunities for the expression of emotion and escape although with more at stake and a different flavour from Glastonbury's rock festival or Wembley's cup final.

Television shown on very large screens has become a feature of pop festivals where the crowd is often so large that they cannot all see the performers as well as they could on television. There are moments at such events when the cameras are turned towards the crowd, who then see themselves caught on screen, many of them using their own cameras and mobile phones to record the event for later viewing on a small screen. Screens are erected in city centres for sports fans to watch at a distance but together. During the soccer World Cup in 2010, Manchester City Council set up an 'official fan park' with three screens at Castlefield with a 17,500 capacity for England matches and two other open-air, public screens in the city also showing matches. Other cities around the country also made provision for these collective televised events that allowed the consumption of alcohol and the sharing of symbols and chants that led to emotional release and created precisely the sorts of events that Durkheim saw as occasions for collective effervescence in which: 'The passions unleashed are so impetuous that they cannot be contained. The ordinary conditions of life are set aside so definitively and so consciously that people feel the need to put themselves above and beyond customary morality' (Durkheim, 2001, p. 163).

At first glance, the collective effervescence and the sense of belonging that comes with embodied co-presence in large social gatherings, are precisely what television threatens to destroy. Instead of going to the festival, the political demonstration or the football match, people may stay at home and watch on television, but as a communicative system it can provide news and information about political and other large social events that then attracts people to go to the venue and participate. It can even begin to provide a focus for a new form of gathering as television screens in pubs or town squares show sports events that fans can watch together. But even if television does not draw the body to an event, it does enable a level of participation and engagement in that event that would not otherwise be possible. To be told about an event by someone who was there can give an impression of what happened and what was missed. Seeing it on television, unfolding in real time, gives more than an impression; it makes possible a sense of inclusion and of joining in. The audiovisual detail and the real-time movement allow the viewer to feel something of what it would be to be a co-present observer, to catch the emotion and a sense of excitement from the exuberance and performance of those who are physically present.

Moral events

As Dayan and Katz suggest, media events such as royal weddings and presidential visits have a variety of effects on their viewers; they interrupt the rhythm of ordinary lives, bringing the event into the household and connecting dispersed viewers through an upsurge of feeling that can *'offer moments of "mechanical solidarity"'* (1992, p. 196 – emphasis in original). The viewers in societies that are otherwise 'organic' in their dispersal and fragmentation are brought together by ceremonial events that are shared through the medium with a spirit of communitas; each separate viewer is aware of the mass of viewers that is involved in the same event. The spoken lyrics to Gil Scott-Heron's powerful 1971 anti-television song ends: 'The revolution will not be televised, will not be televised, will not be televised, will not be televised. The revolution will be no re-run brothers; The revolution will be live.' But for many years, revolutionary uprisings have been televised both at home and abroad without their being any less live, despite the reluctance of falling governments to allow such publicity. The Tiananmen Square protests of 1989 were televised across the world, triggering an international response, as was the 'orange revolution' in the Ukraine in 2004 and 2005. The televised mediation of these events amplified and encouraged local participation, but it also engaged those in the west to become interested in the political events in another country in a more direct way than any newspaper reports. The events of the Arab spring in Tunisia, Egypt and Libya during the early months of 2011 have been heralded as showing the significance of the Internet and mobile phone technology, but the television images have been central to the engagement of the international public in these events.

Robert Wuthnow (1987) cites the 'moral event' of the screening in 1978 of a nine-and-a-half-hour adaptation of Gerald Green's novel *Holocaust* as a ritual that contributed to the moral order in the United States. Apparently, nearly two-thirds of the adult population watched the programme (more than voted in the 1976 or 1980 presidential elections), which attracted a great deal of public commentary. It was 'a symbolic-expressive event that communicated something about social relations in a relatively dramatic way...dealt with a theme of grave moral importance and provoked interest ranging well beyond the series itself' (Wuthnow, 1987, p. 125). For Wuthnow, this televisual event triggered a similar emotional response in many Americans and changed attitudes and confirmed collective values. It provided a stimulus to the collective conscience, but, he argues, also derived its meaning from a period of serious doubt in the quality of American values and trust in its institutions. The programme '...was a ritual event dramatizing the evil of social and moral chaos...The Holocaust in short was a symbol for contemporary chaos, as well as a reminder of historic evil' (1987, p. 129).

The televisual recording of a public event can also make its detail available to the viewer in a way that it is not to the vast majority of those who are co-present. At a demonstration against the gathering of G20 leaders in London on the evening of 1 April 2009, Ian Tomlinson, who was finding his way home from work, was pushed from behind by a police officer, and the incident was filmed by a New York fund manager. The digital video footage shows that Mr. Tomlinson was walking away from the police with his hands in his pockets when, as the result of being pushed by the officer, he fell forward heavily onto the pavement. Despite being helped to his feet by bystanders and supported as he moved away from the scene, he died a few minutes later. The video footage was broadcast on television news and on the website of the *Guardian* newspaper. As well as being available on the BBC website, a copy of the video was also posted on the YouTube website.[1] Contradictions in pathologists reports meant that the police officer involved was disciplined rather than prosecuted, but the public reaction to the televising of the incident has changed police tactics at such events.

As well as connecting people to events that transcend the profane world of everyday life, television creates events in which a studio audience acts as a symbolic collective that draws in those watching at home. Game shows and contests use a studio audience and televisual techniques such as a glamorously-lit set and computer graphics to create a heightened sense of excitement. The game play becomes ritualised with musical themes and repeated words (for example, *Mastermind*, BBC One between 1972–1997; *Weakest Link,* BBC Two, 2000-present). Viewers can join in at home, trying to answer questions, supporting one or more of the contestants, and their shared participation in the game is not merely entertainment or recreation. As Durkheim recognised, recreation and games also function as collective rituals, and it 'is through them that the group affirms and maintains itself' (2001, p. 284). Television shows such as *I'm a Celebrity Get Me Out of Here* (ITV1), *Strictly Come Dancing* (BBC One) and *The X Factor* (ITV1) have participants who either are, or become, minor celebrities as the audience engage by voting on how they perform. Discussion may be with others co-present during the performance, it may be with workmates the next day or it may be through digital media. The progress of competitive shows such as these in which celebrities are vying for public approval is reported in newspapers and magazines with photographs, but is also increasingly a topic for other media. Texting, email, 'tweeting', blogs and social media extend the possibilities for sharing feelings and excitement about the unfolding competition during and shortly after the broadcast amongst friends and other people who follow the show. Television companies use online resources to stimulate this second-order participation – for example, on the *Strictly Come Dancing* website run by the BBC, recent 'tweets' are posted, along with quizzes, photographs, video clips and background information about the show and the personas of the celebrities and dancers.

Folkways and mores

Television can include viewers in the collective effervescence of events that create social solidarity, and it can create moral events of its own such as *Strictly Come Dancing* (BBC One) and *The X Factor* (ITV1), whose rituals become a 'can't miss' occasion in some viewers' week. Moral solidarity is perhaps even more effectively communicated by the representation of ordinary social life. An American contemporary of Durkheim, the sociologist Graham Sumner (1906) was more interested in the routine aspects of culture through which people lived their lives according to a set of unwritten rules or guidelines that he called 'folkways' and 'mores'. If Durkheim saw moral order emerging from sacred and ritualised events, Sumner saw it in the profane, humdrum, everyday practices of ordinary life. Rather than trying to identify the elementary social forms that produce moral culture, Sumner emphasised the variety of ways in which human cultures are formed. His work gathered together and contrasted the variety of folkways and mores in different societies, historically and geographically, to show that moral order is not determined by human biology or even by a superordinate social system such as a religion.

What Sumner was referring to by the term 'folkways' were the routines of habit and skill that emerge as 'expedient' in that they are effective ways of acting to get things done. Nonetheless, they are distinctive of a culture insofar as they emerge through tradition by a process of imitation reinforced by the authority of the elders and ancestors. Folkways are not the result of forethought or reflection – Sumner describes them as 'like products of natural forces which men unconsciously set in operation, or they are like the instinctive ways of animals...' (Sumner, 1906, p. 4). If the folkways were the habits that were followed, more or less unconsciously, the 'mores' were folkways with the added component of philosophical or ethical generalisations about the consequences for societal welfare. He points out that mores were, in the main, constraints against patterns of action that would be discouraged because of some general harm that would result. Morals and mores evolved largely from within the cultural practices of a society and in response to its particular circumstances, including the size of population, the status of priests, the distinctions amongst classes and the amount of available food. Sumner charted, for example, the variations in the moral status of sexual harlotry, polygamy, infanticide and incest as well as the less contentious variations in diet, techniques of work and styles of dress. What can be included in folkways extends to just about any type of regular or routine behaviour; the ways of procuring food (Sumner even discusses techniques for fishing), the uses of language and money, how labour and work are regarded and the standards of living that are appropriate and traditional for a subgroup. Sixty years before Pierre Bourdieu (1984) was discussing habitus in terms of the taste dispositions of different classes, Sumner recognised that variations in diet, dress and style of dwelling were part of the folkways that distinguished cultural groups and that were largely passed on as tradition and habit. Mores give the individual:

> ... his outfit of ideas, faiths, and tastes, and lead him into prescribed mental processes. They bring to him codes of action, standards, and rules of ethics. They have a model of the man-as-he-should-be to which they mold him, in spite of himself and without his knowledge. (Sumner, 1906, p. 174)

The theme of cultural variability is one that generates television programmes that demonstrate how different 'exotic' cultures are from 'our' own. (The scare quotes are to indicate that these terms are always relative and depend on the content of any given programme and its audience.) The form that such shows take is usually to have a televisual guide, a character who speaks the language of the audience and travels to 'other' countries and cultures and experiences the folkways and mores that they find there on behalf of the audience. For example, in the 1980s, the BBC ran a series of *Great Railway Journeys of the World* in which familiar presenters (including Ludovic Kennedy, Michael Wood, Victoria Wood) travelled in different parts of the world. One of the presenters, Michael Palin, went on to make a number of series of travel documentaries using his personality to engage with local people, eating and drinking their foods, living alongside them and enjoying their entertainments and local customs. Another format that emerged was 'expeditions' to more exotic cultures in which Benedict Allen (*The Edge of Blue Heaven* BBC, 1998; *Last of the Medicine Men* BBC, 1999) and, later, Bruce Parry (*Extreme Lives* BBC, 2002, *Tribe* BBC/Discovery, 2006, *Amazon* BBC, 2008) learned about indigenous cultures by living with the people. *Around the World in 80 Faiths* (BBC, 2008–09) was a travelogue providing ethological illustration of a wide variety of religious and ordinary lifestyles, ideas and mores, ritual practices and customs, that explored, through the personality of the presenter, Peter Owen Jones, an Anglican vicar, how different these were from a religious life in the UK. These formats do not claim anthropological rigour but show us how very different mores and material lives can be from that in the comfortably familiar west. More recently, there have been programmes that show us how the folkways and mores of western societies have changed historically, such as Channel 4's *1900 House* (1999) and the BBC's *Victorian Farm* (2009), both of which led to further series and

US showings. Television also provides a continuous socialisation for its viewers into ideas, attitudes and practices of the culture they live in, such as getting the body fit, redecorating or remodelling the house, how to dress, how to cook, where to go on holiday and how to bring up children.

Of course, Sumner was writing before the advent of television about the customs and practices, the 'notions of propriety, decency, chastity, politeness, order, duty, right, rights, discipline, respect, reverence, cooperation and fellowship' that make up the folkways and mores of a moral order (1906, p. 231). If Durkheim wanted to warn society of the risks of modernisation, Sumner saw the variety of social forms as determined beyond the influence of ideas and political process – to interfere would be to risk undermining the evolutionary social logic. The folkways and mores are passed on through habit, imitation and tradition rather than through any particular institutional means but are not effectively prescribed by philosophy or reflective thought. The means of communicating the mores to the masses across generations were myths, legends, fables and religious texts, but Sumner was dismissive of serious philosophical attempts to prescribe folkways (1906, p. 175). Interestingly, he recognised the power of graphic pictorial illustration, in medieval art (sculpture and stained glass), in woodcuts illustrating the Bible and images in daily newspapers. Visual media, especially the contemporary moving images of television, communicate folkways and mores by *showing* them rather than by explaining or describing them. The mundane content of the moral order, the mores and common practices of life become habitual. As William James put it, '...we find ourselves automatically prompted to *think, feel* or *do* what we have been accustomed to think, feel or do, under like circumstances, without any consciously formed *purpose*, or anticipation of results' (1890, p. 24). Folkways and mores are habitual rather than thought or intended, and a mimetic process such as television can reinforce the appropriateness of what we *should* do, but also show what are *not* our customs, our ordinary ways of living.

Sumner is vague about how changes in folkways and mores come about, and reluctant to lay any emphasis on reflective imagination. But John Dewey (1960; 2002), while broadly accepting Sumner's account of group morality, was more interested in understanding how habits change. The origin of the moral response in the individual cannot be traced to instinct, to direct experience, to the abstract ideas of philosophers or religious teachers, but neither is it a cultural pattern that is simply learnt and adopted. Although habits may appear to be repeated actions accomplished without thought, Dewey argues that they are in fact the expression of will, of the desires and means by which human action is achieved. In a reverse of the notion that cognition precedes and determines action, Dewey writes: 'The act must come before the thought, and a habit before an ability to evoke the thought at will' (2002, p. 30). This primacy given to habit applies to sensations as well: 'The medium of habit filters all the material that reaches our perception and thought' (2002, p. 32). Thought and perceptual experience can bring about change, but they must always be based on what is already established as habitual. The television viewer brings to their viewing a set of folkways and mores, habits of living, routines of everyday life, with which they make sense of what they see. Far from being persuaded to simply imitate what they see, the various different ways of living shown are most likely to confirm the rightness of their current habits and ways of being. Dewey's analysis of habit helps to explain the connection between Sumner's folkways and mores; the mores which are articulated reflections are as much determined by the folkways, the habitual sequences of action by which things get done, as the other way around. Changing habits, Dewey tells us, requires pragmatically working out the means by which ends are achieved through a piece-meal, often tangential, application of thought to habits. Simple injunction from the pulpit, the soapbox, the front of the classroom – or indeed the television screen – seems unlikely to achieve change in the morality of society. For habits to be changed, they need first to be reviewed and reflected on, then altered through a stepwise process that is oriented to pragmatically grasped ends.

Diversity

Both Durkheim's and Sumner's early sociology see variations in moral order, such as the size of the group that determines 'moral density', as marking differences *between* societies. However, both seem to overlook the variation in morals and mores *within* a given society. But Morris Ginsberg (1956) recognised the effect of the Second World War on moral diversity within as well as amongst nation-states. Large numbers of soldiers from different backgrounds were brought together to fight alongside each other in the face of a common enemy. The coming together of widely different political and cultural groups as the 'Allies' against the equally culturally diverse 'Axis' powers of the Tripartite Pact of 1940 (Germany, Italy and Japan) blurred moral cultures and challenged the idea of a single moral order connected to each nation-state society. There was a universalising of moral sensibility to the shared response of disgust and outrage about the Holocaust, but Ginsberg also pointed to the moral diversity amongst and also *within* nation-states. Neither, he argued, was a result of the 'moral bewilderment' created by the Nazi regime but was part of the normal process of social evolution and belief systems like Christianity could be used to support very different moral positions. Even within Nazi Germany, some people were moral nihilists who indoctrinated others, some were fanatics, and yet others were simply loyal citizens, used to following authority despite their moral bewilderment (Ginsberg, 1956, p. 3). He also pointed to the moral diversity in post-war Britain despite the recent unity against a common enemy; strongly-held differing views on capital punishment, the indissolubility of marriage and birth control, for example, were not reducible to a single moral order or an overarching system of moral principles (Ginsberg, 1956, p. 100). All societies may have morality, but that does not mean there is a single moral order that includes everyone. Different social groups with different experiences and knowledge may have different values and practices; Ginsberg gives the example of different religions with different moral teachings having similar values, while varying knowledge about the effects of hygiene or birth control will lead to different values within religious groups. Even the responses to swearing will differ according to social group and context.

The moral values and habits of custom vary amongst social groups and affect how they engage with what is shown on television. Early television presumed a unified culture that was upper middle class (public school accents and evening wear for presenters), but since the 1960s, broadcast content has recognised diverse class and cultural origins and interests. Soap operas such as *Coronation Street* (ITV, 1960 till the present) and *EastEnders* (BBC, 1985 to the present) and 'kitchen sink dramas' (for example, *The Wednesday Play* BBC, 1964–70) have self-consciously reflected the lives of working-class communities, and in a wide variety of television shows, class boundaries are linked to profession (criminals vs police vs lawyers; doctors vs nurses) and so to different interests and values. Some programmes presume to address the values of a gender group (for example, *Top Gear* BBC, 2002 vs *How to Look Good Naked* Channel 4, 2006), although not being of the gender group and even not sharing the values is no barrier to finding some pleasure in watching. Children are addressed differently through their own programmes and channels (CBBC – UK; Disney Channel – US), and in the UK the 'watershed' and the Broadcasting Code puts a responsibility on broadcasters to protect the under-eighteens from material that might impair their 'physical, mental or moral development'.[2]

Producers of television attempt to respond to the diversity of the audience they are addressing, and the diversity of that audience produces different responses to the content of programmes (Morley, 1980). Bev Skeggs (Skeggs et al., 2005; Skeggs, 2008) and her research team have recently shown, for example, that while participation in reality television was seen as 'morally bad and exploitative' by audience members in a middle-class focus group, a black and white working-class group saw the same programmes as showing 'an imagined possibility of a less constricted future' (Skeggs et al., 2005, p. 19). A third, British Asian, group enjoyed the sociality of watching reality television together, but appearing on the show would be something shameful for them. Diversity in socio-cultural situation, in life experience and in educational level, leads to differences in the moral meanings for different groups of viewers. The sorts of moral issues that Ginsberg mentions as being characterised by diversity – swearing, marriage, adultery, reproduction, the treatment of children, sexual behaviour – are all themes that regularly make up the narrative content of television programmes. A variety of behaviours and their consequences are displayed that assume only the broadest level of moral consensus on what is acceptable. Within a secular moral order that presumes individual behaviour should be oriented to minimise harm to others, there is room for great variation in the outcome and impact of different moral perspectives. The audience has the opportunity to explore how values that they do not share, play out through the vicarious experiences of those real and fictitious characters they see on the screen.

In the post-war period, Ginsberg was interested in the possibility of a unified moral order around concepts of justice and universal human rights (he was involved in drafting the UNESCO statement on the Race Question in 1950 – a response to the racist ideology of Nazism). But on less vital issues, Ginsberg adopts a form of cultural relativism through which he argues persuasively that different feelings and impulses are appropriately part of the response to moral questions. Consistency in the moral order can act as a constraint on individual wishes and desires, but mediated culture including television, what I have called 'mediated solidarity' (see above, p. 50), exposes people to a greater range of moral systems and situations than they would ever experience in their everyday lives. This can lead to tolerance of difference but may also encourage reflection and discussion; what is shown on television may enable much more rapid shifts in the folkways, mores and the moral order of subgroups and ultimately the whole society.

Drawing on sources including Sumner and Ginsberg, Steven Lukes (2010; 2011) argues that rather than thinking of a moral order in the singular, we must recognise a diversity of *moralities*, some of which may be characteristic of particular societies in different parts of the world at different times, others of which may exist alongside each other in the same society at the same time. Perhaps the human capacity for morality can be traced back to the cooperative strategies and emotional commitments amongst other species (de Waal, 2006), but there is a step change in human societies from instinct to abstract principles, obligations, values and motivation 'by an ought' (Korsgaard, 2006). So, morality may be a distinctively human capacity, not one that determines the content of moral order but does suggest an origin in the relations of interaction; how to show gratitude or offence, goodwill or resentment, love or hurt. Being able to show these reactions is reciprocal, so people are able to recognise the same attitudes in others as evidence of underlying emotional states. Rather than having to tie morality down to a specific set of values and behaviours, Lukes is able to argue that the analysis of a given set of reactive attitudes, the 'repertoire of moral sentiments that are realized in strikingly different ways in different social contexts' (2010, p. 558), indicates the underlying morality that has been constructed by that society at that time. Lukes nicely distinguishes amongst different types of concern and orientation: ' ... we *proscribe*, say, torture and rape but we are reluctant to *prescribe* how people ought to act and live their lives' (2008, p. 137). That there is a diversity of values and moral norms he takes as a fact, but this does not necessarily lead to moral relativism or shying away from the possibility of universal human rights (Lukes, 2008, p. 136).

Interaction

The moral order of interaction operates at a micro-level compared with Durkheim's account of social solidarity, but both recognise that the existence of human society is dependent on morality. The moral order of a society is not imposed or legislated for but emerges out of the acceptable patterns of interactional behaviour. Mutual recognition of shared morals, customs and laws enables interactions between people to be based on reasonable expectations of each other (Luckmann, 2002, p. 28). Where strangers meet, they may be cautious but presume moral standards of interaction are shared until it is evident they are not. Sometimes the setting (for example, a battlefield, a dark street in a rough part of town) prepares strangers for a significant difference in values (for example, in respect for other peoples' lives). Interaction here includes communication through speech and gesture, but all bodily action and practices, such as ways of walking and demeanour, are culturally meaningful. Turowetz and Maynard distinguish between the moral order *of* interaction, the assumptions about what is normal and natural, and the moral order *in* interaction, when people 'justify, impugn, excuse, praise, exemplify, exculpate' actions (2010, p. 504). The appropriateness of actions can become the topic of communication when it is commented on with approval or approbation. For Anne Warfield Rawls, moral order in interaction is a 'constitutive order', the tacit level of agreement about behaviour and action that is necessary for mutual intelligibility and indeed for society to exist which, unlike institutional order, lacks a written code or referee and has a routine, unnoticed character (2010, p. 95).

Erving Goffman (1961; 1972) has described the ritual and ceremonial features of human interaction that are often expressive of relative social status, a feature of the moral order in which the worth of people and the appropriate modes of action for them are established through practices that shape their 'moral career'. Goffman describes, for example, the ceremonial order of ordinary interactions through which individuals are 'given face' (1972). Sometimes the ceremonial effect of interactions is to systematically reduce the status of a person as in the degradation rituals that Goffman observed on the back wards of a mental hospital; taking away the person's clothes, regimentation and strict control of activities (1961). The ceremonial order is the observance of rules of conduct relating to:

> ... matters felt to have secondary or even no significance in their own right, having their primary importance – officially anyway – as a conventionalized means of communication by which the individual expresses his character or conveys his appreciation of the other participants in the situation. (Goffman, 1956, p. 476)

In many television shows – magazine programmes, talk shows, news programmes – there are ceremonies around the arrival and departure of guests, with warm words, applause and thanks. Talk shows such as *The Graham Norton Show* (BBC One, 2007– present), for example, often have an audience present who participate with appropriate clapping, laughter and even cheering. Guests and contestants are greeted with ceremonial hugs or kisses by the host, and they often recognise the audience with nods and waves. These are ritualised performances of emotion for public consumption that reproduce the intimacy of warm and friendly interpersonal encounters. Interactions on these shows can include ritual degradation, usually of a member of the audience or a contestant who is made fun of or humiliated for the pleasure of everyone else. A skilled presenter such as Graham Norton (or Bruce Forsyth on *The Generation Game,* BBC One, 1997–2005) can manage this in a way that draws the victim into the laughter. Sometimes ritual degradation is not for laughter but is seen by everyone, including apparently the victim, as justified by poor performance (as in *Weakest Link,* BBC Two, 2000–2012). On the other hand, ritual degredation such as happens to some of those who are auditioned in the early rounds of *X Factor* (ITV1) is a shock to the contestant. Lacking in stage presence and having very weak singing voices, they nonetheless have got through the pre-auditions with researchers to appear in front of the panel and a television audience of millions. Their humiliation and disappointment must have been predictable for everyone but themselves, which makes all the more painful the laughter and howls of derision they suffer from the audience and the pained expressions on the faces of the judges.

The 'moralities-in-use' of interaction sustain the moral order within groups and milieus and include facial expression, gesture and demeanour as well as the spoken word, according to Thomas Luckmann (2002). The traditional mode of moral communication, he tells us, took the forms of complaints about others, accusations of misdeeds, apologizing for faults, becoming indignant, praising or condemning, pronouncing maxims or proverbs, providing or seeking advice, gossiping, preaching, or swearing (Luckmann, 2002, p. 22). However, he argues that in late modernity this means that beyond local settings there is uncertainty about what the prevailing moral order is, and this leads to 'indirect moralizing' that involves, for example, questioning an action instead of simply judging it. Instead of directly saying, 'You shouldn't do that!' and by saying, 'I don't understand why you did that', the moral appropriateness of the action can be called into question indirectly. Luckmann offers some examples from his empirical research of indirect moralizing that question moral propriety without giving a direct judgement: euphemisms, false starts, reformulations, jocular modulations and prosodic devices, such as a complaining tone with a semantically neutral remark (2002, p. 31). He points out that mass media, including television, lend themselves to 'indirect moralizing' by embedding it in a script or exchange that is indirect for viewers; it is a character or a person on the screen whose moral status is being questioned, not the viewer's. Indirect moralizing allows for the variation in moral sensibilities of the audience, and its lack of explicitness means that it is not so likely to offend or contradict the moral order of different segments of the audience. The viewing public tolerate the indignity that some *X Factor* contestants are exposed to because it was someone else who should have warned them that their dreams of fame were going to be dashed.

Play with morality-in-use on television can reinforce moral values indirectly. For example, in an episode of *The Graham Norton Show* (BBC One 07/03/11), the actress Miriam Margolyes was prompted to tell a story in which, having declared that she is a lesbian, she describes how as a student she stopped her bicycle beside an open car and asked the American soldier in it, 'Would you like to follow me to my college, and I'll suck you off?' The audience squealed with laughter and clapped while Graham Norton laughed with his mouth closed and his faced screwed up in mock disgust, and another guest fanned his face with his hand as if to cool his embarrassment. The third guest made a mock show of getting up and walking off, apparently unable to stay in this outrageous company. Shortly after, Margolyes milked the story by reflecting, 'I should explain; I thought I was a good girl, because a bad girl would have had intercourse'. The story is worth telling because of its account of what is recognised as an immoral act. (Even though unembarrassed in the telling, Margolyes recognises the moral transgression through the hindsight of her explanation.) The reaction of the studio guests and even the peals of high-pitched laughter from the audience serve as indirect moralising in a way that leaves open a range of possible moral positions for the viewer, none of which needs to be expressed. The viewer is in the situation of being a silent participant, but the physically present studio audience helps to mediate the experience of the viewer at home – the smiles, gasps and laughter from the studio audience cue not only each other but also those watching. The camera, however, gives the home viewer the close and intimate sense of participation in the interaction on stage that the guests enjoy.

Television can enable the moral proprieties of interaction to be stretched in a way that is difficult in ordinary settings. *Embarrassing Bodies* (Channel 4, 2007–present) breaches interactional norms that prohibit showing damaged, diseased and deformed parts of the body, especially the 'private parts'. The programme, broadcast at peak mid-evening on Friday evenings, is appealing precisely because it challenges the customary reluctance to talk about illness and body parts. The programme has its own website that includes video clips from various programmes, questions and comments from members of the public and penis and vulva 'galleries' as well as the usual breakdown of episodes. At one level, the programme is a modern 'freak show' (Bogdan, 1988), but the participants are interacting with a doctor offering advice and reassurance, legitimating the transgression on the grounds of public education about the human body.

Postmodern ethics

The sociologist who brought morality onto the agenda of postmodern social theory is Zygmunt Bauman. In a series of works. Bauman (1989; 1992; 1993; 1995; 2002) has explored the morality of high modernity and the potential for a different, postmodern ethics. As we will see in Chapter 7, Bauman is critical of modern television for creating the 'telecity' in which we look but do not care, we watch without responsibility as if we are spectators at a zoo. But here I want to recognise the importance of his concept of postmodern ethics because, contra Bauman, I think that television actually contributes to postmodern morality. He argues that, until the Enlightenment, everyone participated in the same form of morality, which was linked to religion and the law of the land and to the common practices of the society. Everyone treated it as 'natural', and all but the rich and powerful followed the code. They were able to publicly support the values of the accepted code while privately indulging their desires. The Enlightenment changed this unitary moral system by suggesting that it was possible for everyone to be free to fashion themselves and determine their own lives according to their own choices. This led to many ethical systems, each competing to discover the true nature of humankind without being based on religious ideas: 'Doing good had to be shown to be good for those who did it. It had to be desired for the benefits it brings – here, now, in this world. It had to justify itself as the *rational choice* for a person desiring a good life, rational because of the rewards it brings' (Bauman, 1993, p. 27). Philosophers debated the nature of humankind and how it should be and discussed the various modes and principles for right action. The principle of rationality is what makes the ethical systems that emerged *modern*; they were based not on tradition but on reason and interests.

The modernist logic of a systematic and rational ethics led, however, to the lowest point of modernity, the Nazi Holocaust (Bauman, 1989). A system of ethics – a way of living, a way of relating to other people – was established both as a political regime and a cult, as an army and a population were persuaded to apply rational principles to organising genocide. The ethics of Nazism assigned everyone a role and a task in which following the system was more important than the human consequences. The regime didn't tolerate deviation from the role, but it rewarded allegiance and applauded innovation in realising its aims. This was how death camps came to be designed; reason was applied to the industrialisation of death through building gas chambers and setting up bureaucratic systems capable of identifying, counting and shipping those to be killed. Individuals did not feel responsible or morally accountable for their actions, they simply followed orders without resisting or thinking, a process that Bauman (1989) calls 'adiaphorization'. (Tester helpfully summarizes its meaning as the 'social construction of indifference' – 1997, p. 78n.) This is an important concept for Bauman's critique of the ethics of modernity, the severance of cruel acts from moral guilt, and he accuses television of having extended the power of bureaucratic adiaphorization by what he calls the 'insensitivization' of viewers to human suffering and the distance over which violence can be so easily perpetrated. It is television that is largely responsible for what he calls the 'staged cruelty' in which violence becomes routinely introduced to daily life:

> There is hardly a day without dozens of corpses and killings being brought into our view on the television screen, whether in the time-slots described as 'the news', or such as are classified as drama or comedy or police series, feature films or children's programmes. (Bauman, 1995, p. 149)

The electronics of modern warfare mean that so much actual violence is realised by looking at screens and pressing buttons that those doing it are as distant from the moral significance of their actions as if they were watching television or playing a video game. But there is in Bauman's argument a dangerous tendency to generalise about the effects of modern technology; there is no reason to believe that a computer weapons system leads to less of a sense of moral responsibility than a knife, rifle or machine gun. Neither does a nightly diet of violence on television turn people into Nazis or mean that they lose any sense of moral responsibility.

The attempt to refine a unitary, systematic ethical system that Bauman associates with the instrumental rationality of modernity can, he hopes, be countered in the postmodern era by a different approach to morality. Postmodern society is: ' ... a self-reproducing, pragmatically self-sustainable and logically self-contained social condition defined by *distinctive features of its own'*, it is 'modernity emancipated from false consciousness' (Bauman, 1992, p. 188 – emphasis in original). The false consciousness of modernity was the belief in the capacity of instrumental rationality to focus and direct society, and what postmodernity brings is the possibility of autonomy for individuals so that: '…moral selves do not 'discover' ethical foundations, but…build them up while they build up themselves' (Bauman, 1995, pp. 19–20).

Bauman has a particular take on the sociological features of a postmodern morality in which individuals recognise their responsibility by responding to the situation rather than simply following rules laid down by someone else. This is a more difficult task for the post-modern individual, and Bauman discusses the complex criteria by which moral decisions may be arrived at; of central importance is the proximal experience of intersubjectivity, a relationship with the Other that makes the person into a subject. It is from Levinas that Bauman derives this essence of humanity in the moral responsibility for another human being that is not dependent on reciprocity. This is a total responsibility, an unquestioned and unconditional responsibility that includes interpreting and understanding the needs of the Other, a relationship that is to do with closeness and touch (Bauman 1993, pp. 85–108).

This is a very particular approach to morality based on the caress and the proximity of an Other, which seems to turn away from the sociological dimension of morality as a relationship with those who are beyond immediate interaction.[3] Bauman takes the 'moral party of two' as the origin of moral responsibility, and although he explores how it might extend to those who are within sight and hearing rather than touch, the problem is that he finds the moral impulse of humanity becoming ever more diluted as it becomes systematic and able to consider excluding some others from the sphere of moral responsibility. Postmodernity, Bauman seems to believe, indicates a running down of the trajectory of modernity towards ever greater distance between people and a return to a messier form of society. He sets out the marks of the moral condition from the postmodern perspective: the moral ambivalence of human beings, moral phenomena as not rule-guided, morality as incurably 'aporetic' or uncertain, morality as not universalizable, morality as irrational, moral responsibility *for* the Other before one can *be* with the Other (Bauman, 1993, pp. 11–13).

Bauman's critique of the ethics of modernity and his understanding of post-modern morality are too complex to consider or question in detail here, but I want to argue that his Levinasian account of morality is profoundly unsociological; it offers a humanistic essentialism based on the dyad but fails to adequately account for the complexity of the moral order that enables societies to exist. His sentimental concern with proximity and fear of distance overlooks the need for care to transcend relations of touch if societies are to thrive. What Bauman dismissively calls 'telemediation' (1993, p. 178) can also be a channel of interest and concern that can lead to emotion and to caring action. In fact, the multiplicity of perspectives offered on television, which fail to be cohered by any rules, are communicated through an audiovisual medium that lends itself to emotion more than rationality. Far from generating a universal morality, what television does is contribute to a messy postmodern morality in which viewers are left to make their own judgements. Human beings are intrinsically morally ambivalent, but their relationships with others, both in proximal experience *and* through mediated intersubjective experience, can stimulate a sense of responsibility towards and care for the Other.

Conclusions

What I have argued in this chapter is that morality is tied up with the very thing that society is – it is meaningless to have a society without morality, or morality without society. Morality involves the obligations one feels towards those people one does not 'know', who are not relatives, neighbours or others with the shared interests of a community but can be recognised, nonetheless, as members of the same society. Philosophers have attempted to identify universal principles from which morality could be deduced, but with Emile Durkheim we can see that the moral order of a society is apparent in the various cultural forms that prescribe approved behaviour. He sees the individual person deriving her or his morality from the society in which they live and not from any introspection or discovering of a true or universal morality. It is in laws, customs, habits, mores, routines and interactional practices that a society sets out its moral order; it becomes manifest in the proscription of behaviour and in the sanctions attracted by transgressions. Its various forms may be written about – even specified in writing, as is a legal code – and they may be formally taught through the family, church or school.

The moral order of a society leads to its social solidarity and is strengthened by the capacity for people to come together emotionally through processes of 'collective effervescence' and to express the shared values of the *conscience collective*. Media events provide an opportunity for a 'coming together' of people who are not co-present, and I will argue later (in Chapter 8) that we can see such expressions of the moral order as leading to an 'imagined community' or a 'social imaginary'. Here, though, it is enough to note that television contributes to a 'mediated solidarity' that is characteristic of those cultures where one-to-many electronic media are available to most people.

If Durkheim emphasises the macro-social process of moral order that leads to cohesion, Graham Sumner emphasises the mundane and everyday features of social life that are consistent within a society but which vary amongst societies, over time and across nation-state boundaries. The folkways and mores, the habitual ways of doing things, are learnt and passed on as if they were natural, but in fact they are formed in relation to the particular society, its size, economy and complexity. It is not through philosophical reasoning that they are sustained but by images that are shared and passed between generations; television provides a contemporary source of such imagery that expands our exposure to representations of different ways of living. Sumner points to the contrast in folkways and mores amongst different societies, often at some distance or time interval, but Morris Ginsberg commented on the coming together in the second half of the 20th century of moral views that crossed between very different modern societies. At the same time, there were identifiable variations in the moral orders subscribed to by different groups within these societies.

Diversity and change in the moral order – or more, properly, moral *orders* – of modern societies are dependent on how they are passed on. These various moral orders are acquired through the continual process of interaction; through living in the social world, imitating and adopting as habits those ways of acting that are acceptable and appropriate. Erving Goffman and Thomas Luckmann, among others, have pointed to the ceremonial practices of interaction that not only sustain a moral order but also enable it to cope with variation and change. The traditional form of instructional 'moral tales' gives way to indirect moralising which becomes more prevalent as the way in which moralities in use are sustained in interaction in late modernity. This mode is particularly noticeable in the interactions shown on television in which the viewer is not expected to make a direct moral judgement but to draw a moral conclusion by inference.

The variability and ambiguity of moral orders could threaten the coherence of late modern societies as mass communication confronts consumers with different morals that might threaten their values and ways of life. But a postmodern ethics has emerged which re-emphasises the responsibility of the individual to make moral judgements and choose which paths to follow. There is a risk of moral relativism, of choosing pragmatically according to the situation and tolerating the behaviour of others on the grounds that their standards are different. But tolerance of the variation in lifestyles is not sufficient reason for abandoning all principles, since there are experiences common to all societies from which at least some foundational standards can be derived.

Television extends, rather than displaces, the direct interactional field for acquiring morality and adds to the ways that people can connect with the moral orders of their society. It can show a range of behaviour and ways of acting, and it can also show the material and interactional consequences of various values behind people's ways of acting. Television can promote social solidarity as it shows familiar patterns of behaviour and responses that are shared, but it can also lead to tolerance and acceptance of behaviour that is different. It can stimulate reflection on current attitudes and values, but it can also be an occasion for judging behaviour that transgresses what should be universal values. Because the small screen is only a medium and not a unitary institution, it cannot present a coherent religious or ethical system. What it offers to its individual viewers is the autonomy to bring their own moral judgements to bear on a wide variety of situations presented in very different ways. Television contributes to a culture in which morality is postmodern in its contingency and its ambiguity, but the small screen enables viewers to contemplate the outcomes of different lines of action as they consider, albeit emotionally rather than rationally, the moral issues raised by the content it shows.

4
Televisuality: Style and the Small Screen

What makes television a distinctive medium? I have suggested that just what 'television' shows us has evolved, and in this chapter I want to point to some of the features that make television different from what it was when if first became a mass medium. John Thornton Caldwell (1995, p. 55) and Jeremy Butler (2010, p. 26) both refer to the traditional form of television in which the medium effaced itself and simply transmitted visual information, as 'zero-degree style', by which they mean the process was conventional and not self-consciously stylised.[1] This was the standard way of capturing live studio shows (music, games, comedy and variety performances, soap opera) as well as reporting outdoor events including sports, in which the camera simply gave the television viewer the perspective of a spectator at the live event. For recorded drama, there was a stylistic element in that close-ups, and a number of camera angles were used, but the aesthetic was restrained and derived from early movies rather than making distinctive use of what television could do. While the cinema had from the beginning developed an aesthetic interest in the look of the screen imagery that connected with the story of the film, television as a technology was more constrained and directed towards what was live and actual. Things changed as television developed its own distinctive 'look', and I will borrow Caldwell's term 'televisuality' to refer to what has become the aesthetic of television as communication, although my use of the term is rather broader than his. For Butler, as we will see, the evolution of a televisual aesthetic is primarily about style, but Caldwell sets the emergence of a distinctive televisual aesthetic against the competition for an audience with cinema and the technological changes in cameras and recording equipment. Television is, after all, an industry in which economic capital is invested and managed with an expectation of a return, either in financial terms or as a public service. To organise the technical equipment and teams of skilled people requires a corporate institutional structure to plan and steer the making and broadcasting of programmes. Unlike one-to-one communication in which the two parties shape communication iteratively according to each of their interests and perspectives, one-to-many communication operates on ideas generated within a corporation with very cumbersome feedback systems of public criticism through other media, individual complaints and comments and the responses of advertisers and regulators, but above all, the information from viewing figures.[2] In the early days, the institutional structure was directed toward organising and creating largely live programme content to fill the schedules; things had to be made quickly and did not need to last. But as the technology changed, and the economics of television became more competitive, aesthetics became as important for the small as for the big screen.

At first glance, the development of a televisual aesthetic has nothing to do with the capacity of the medium as a contributor to the moral imaginary, but I want to argue that there were consequences of tele-visuality for morality. The development of a distinctive televisual aesthetic attracted audiences and engaged them with the pleasure of simply viewing what was happening on the small screen, and as it did so, the emphasis was taken away from narrative content. Television programmes could be watchable even if the storylines did not achieve a very satisfactory closure, so they became more open and ambiguous, stimulating imaginative work by the viewer to make up their own sense of what was happening. Instead of characters having to be clearly 'good' or 'bad', it became easier for them to depict shades of good or bad; improved definition allowed more subtle camera work that revealed greater complexity in the portrayal of people, their actions and their emotions. At the same time, viewers could stand back from the action through the screen and simply enjoy the changing images or the way they were constructed. This greater range of possibilities meant that audiences might enjoy what they did not agree with; they would tolerate characters and storylines that breached moral norms...because the tele-visual image looked good. This potential for ambiguity, both in the depiction of the consequences and outcomes of lines of possible action, and in the viewer's engagement with what was on the screen, created an ambivalence that contributed to a postmodern moral sensibility. The dynamic imagery that was being added to the moral imaginary became more subtle, attractive and memorable while at the same time being more ambiguous in outcome.

What television became grew from its early relationship with cinema, so that is where I will begin. Then I will look at televisuality before revisiting Raymond Williams's concept of 'flow', a distinctive aspect of televisuality that is not considered by either John Thornton Caldwell or Jeremy Butler.

From film to television

Traditional cinema film works by the projection of light through a series of frames of partially occluded plastic material (originally celluloid but nowadays polyester) and then through a lens from which a cone of light falls onto the screen, casting an image many times the size of that on the film. In the cinema, this image is projected from behind and over the heads of the audience who sit in rows, in near darkness, facing forwards, their eyes focused on the moving image on the screen. The audience is made up of strangers in quite large numbers who have come together purely for the purpose of sharing the performance. Each has made the effort to get to the cinema and will have paid and waited or queued to see the feature film. Filmgoers sitting in the dark in a cinema often object to other people talking during a film, leaving the projected images and sound not only without any competition but also other interpretation. The co-presence of others may intrude through coughs and sneezes, noisy sweetwrappers or the sound of sucking on straws, but their laughter, delight or nervous reaction to the film may make watching into a more social experience.

The cinema is a film *theatre* in which the performance on the screen is the focus of everyone's attention, and for film theorists it is the *gaze* that is characteristic of how the moving images are consumed. John Ellis famously distinguished the *gaze* as the mode of spectatorship in the cinema from the momentary and casual *glance* as the mode for television watching:

> Gazing at someone is a slightly uncomfortable experience for the person looked at, and a sign of intensity of attention on the part of the person seen. Gazing is the constitutive activity of the cinema. Broadcast TV demands a rather different kind of looking: that of the glance. Gazing at the TV is a sign of intensity of attention that is usually considered slightly inappropriate to the medium. Its most active proponents are children, watching despite their domestic circumstances, learning the ways of the world and its narratives. (Ellis, 1982, p. 50)

This distinction in the modes of looking appropriate for the two media is linked to the size of the screen as well as the material and social environment. The cinema screen is big and can fill the visual field, while the television screen is small and fills only a fraction of the visual field. Sitting closer to the TV screen – as children sometimes choose to do – can increase the immersive effect on the visual field of the watcher, but domestic television viewing tends to have a higher ambient light so that other things can be noticed or watched. Unlike the theatre- or cinemagoer, television watchers are usually at home, in well-known surroundings with, if any, a few familiar people in comfortable seating arranged around a small screen that emanates light to make its image. The ambient sound and lighting of television watching are domestic – the doorbell, the telephone, the kettle, the shouts of family members elsewhere can all be heard, and there is usually sufficient light for those who wish to read or gaze at other items of visual interest (for example, windows onto a garden, pictures on the wall, flower displays). This different context of viewing, combined with the continuous flow of connected content rather than the distinct and simple programme of trailers, advertisements and feature film in the cinema, means that the small screen is a rather different sort of phenomenon.

What made television into the system that became a mass medium in the 1950s was, Brian Winston (1985) argues, linked to the earlier cultural impact of cinema and radio on modernity as popular entertainment. These media had been built on 'the addiction to realism in the culture, the supremacy of the nuclear family and its home, the industrialisation of the entertainment business' and it was 'spare industrial capacity at the end of the second world war…and the push to consumerism' that produced television as we now know it (Winston, 1998, p. 110). To an important degree, what appears on the screen is constrained by the technological apparatus for gathering the image and presenting it, and television has always been judged against film. In the early days of tele-vision, cameras were large and heavy and needed good lighting; they worked best in a studio. However, because the light emitted from film made it easy to scan, the capacity to broadcast film was incorporated into the various competing forms of television as they were built and demon-strated (Winston, 1998, pp. 95–110). As television was being developed and tested, particularly between the two world wars by different people and companies in different countries, the definition of each system's image was judged against that of film, initially 16mm amateur film.

It was the lower definitional quality of television as against film that was behind Marshall McLuhan's (1994) distinction between 'hot' and 'cold' media. Television was a 'cool' medium because the small, low-definition image made for a high level of participation by viewers as they interpreted what was going on. At the time that he was writing, in 1964, television's black-and-white image was made up of 405 interlaced lines of flickering light as a thin beam whizzed across a cathode ray tube, whose convex surface and rounded corners meant that the edges were less defined than the centre. In contrast, the 35mm cinema projected movie was in high-definition colour on a rectangular screen with fuller, stereophonic sound – it left much less for the viewer to do. Electronic technology, McLuhan argued, had begun to create boredom and to produce concern with *effect* rather than *meaning* (1994, p. 26), and as entertainment, media were '... radically upsetting... as the cool TV medium has proved to be for our high literacy world' (1994, p. 31). Although literature is still today treated as a 'higher' cultural form than television, the exchange between the small screen and high culture has increased to the point that there is no longer any upset.

McLuhan's categorisation into 'hot' and 'cool' media was always provocative and contentious, and it has become even more so as television technology has developed through a series of innovations – increased number of lines, colour, flat square screens, digital cameras, digital transmission and display, high-definition, non-interlaced, 3D – so that it is no longer a low-definition medium. But is increased definition enough to make the small screen a 'hot' medium? Television still works differently from film projected onto a screen, and this continues to affect the way it is viewed:

> The TV image is not a still shot. It is not photo in any sense, but a ceaselessly forming contour of things limned by the scanning-finger. The resulting plastic contour appears by light through, not on, and the image so formed has the quality of sculpture and icon, rather than of picture. (McLuhan, 1994, p. 313)

Despite the increased refresh rates and disappearance of lines into pixels, the 'scanning finger' of light still emanates from the screen whether it is a cathode ray tube or a liquid crystal or plasma display panel. Television images continue to be sculpted out of light rather than being formed by still photographs shown at sufficient speed to convince the human brain that it shows movement. McLuhan wrote of television as a 'mosaic' that required a synesthesic combination of tactile and kinetic senses in the viewer to close the mesh, the gaps among the pieces of information provided through the visual channel.

Of course, his central argument was that 'the medium is the message', so it is not surprising that he emphasised the particular characteristics of the medium as a technology that extended human capacities rather than taking much notice of its content.

Live television could be distinguished from film precisely because some techniques – montage, slow motion, time lapse, stop trick – are derived from film as a series of still takes that, the historian Friedrich Kittler tells us, make it a medium of abstraction from the real: 'As phantasms of our deluded eyes, cuts reproduce the continuities and regularities of motion...' (1999, p. 119). Digital technology, however, has been able to reproduce all of these aesthetic tricks and even insert them into the flow of live television (as happens with the slowed down 'action replay' in sport). Technical differences also produce a 'look' that used to distinguish studio cameras from video recordings and video from film; film has traditionally had a warmer, slower, softer, fuller effect with a granularity and palette dictated by the film stock and lenses with a narrow depth of field that produce greater contrasts in light, while video had the reputation of being colder, sharper and harder, often shot with a wider depth of field and interlaced as it was recorded. Film has always been used to make programmes for television and broadcast alongside live programming and video recordings. However, as high-definition digital cameras and screens have replaced the analogue technology of tube-based cameras and videotape, the technical distinction between 'film' and 'television' media has all but disappeared. Digital video techniques are already used in the making of both movies and television, and the skills and equipment can often be interchangeable even though styles and standards may be different. Ultra-high-definition cameras will mean that digital recording has at least as much information as film and will be able to produce very-high-definition moving images for small as well as big screens.

If there is a difference between the cinema and TV spectatorship, we might expect there to be differences in the visual style – and the argument about televisuality is precisely that there is such a difference. John Ellis argued that television's 'regime of visualisation', characterised by the glance, stressed immediacy, visual intimacy and co-presence, focusing on what was significant rather than on detail (1982, p. 137). The poverty of the small screen image meant that it was dependent on sound to ensure continuity of attention – the sound could be heard from anywhere in the room and even beyond, carrying the key information for news, documentary and comedy programmes. For Ellis, this meant that people 'looked in' on television, which was 'gestural' in its intercut, multiple-camera view of things, in contrast to the fetishistic voyeurism of the cinema screen. This distinction in spectator mode is difficult to buy into thirty years on (see Carroll, 2003, pp. 265–79). Films made for the cinema work remarkably well on the small screen and constitute a high proportion of broadcast output and home video watching. Furthermore, many made-for-television programmes, such as serials and dramas, use very similar storytelling techniques to the cinema and often last about the same length of time, requiring the same level of attention. As the technology of the television has improved (widescreen, realistic colour, high definition), it has developed the capacity to show the same level of detail as a cinema film so that distinguishing between the perceptual demands and viewing pleasures of the cinema and the small screen has become more difficult. Some may only glance at the television (as they read a newspaper, talk on the 'phone, do the ironing or check emails), but some will gaze at it much as they would in a cinema.

Televisuality

Closed-circuit television – as is generated by a security system, for example – simply captures what is in front of the camera and often fully records it in real time, but television that is produced for distribution and viewing, even at its most basic, has its own style. The techniques of arranging camera angle, distance and framing in relation to what is captured are basic to photography, film and television camera work but shape what is shown in ways that are often not apparent to the viewer for whom what is recorded seems to be the only possible image. Since the early days of large, difficult-to-move television cameras stuck in a studio under powerful lights in front of which presenters performed, technological developments have reduced the size of the standard camera to something that can be held and moved by hand or by electronically controlled devices in ways that can both mimic the view of a person and go very much further than any humanly-embodied view. Nowadays, the 'flying cameras' over a football pitch can take the viewer in a moment from the player's perspective to a bird's-eye view of the whole arena.

The 1980s and 1990s were a period of particularly rapid technological development in videography that transformed the making of television. Instead of using film where possible and emulating the 'filmic' style of full-feature production movies, television began to develop a distinctive style of its own that revelled in the possibilities available to lightweight cameras and video editing. John Thornton Caldwell dubbed this style 'televisuality' to emphasise its emerging 'aesthetic based on an extreme self-consciousness of style' (1995, p. 4), which, at least to begin with, was impossible to emulate with the heavy 35mm camera equipment of the movies. Exhibitionistic, excessive, self-consciously stylised, televisuality is firstly a performance in which the television image demands attention through effects created by the camera and video editing. Televisuality plays on movement – of the camera, of graphics, of images – that create dynamism, activity and energy on the screen to stimulate and engage the audience. Caldwell writes, '...this self-consciousness of style became so great that it can more accurately be described as an activity – a performance of style – rather than as a particular look. Television has come to flaunt and display style' (1995, p. 5). Caldwell argued that Ellis's 'glance' theory was outdated and that televisuality was precisely about attracting and holding the viewer's gaze. In rejecting the dualism between film and television, he argued for a return to the televisual apparatus, the technologies and the style used to make what gets shown on the small screen.

There were conventions of television that were developed around its form in the early days; the low-quality image, the need to fill the schedules and the need for most television to be broadcast live. But by the 1980s, the technology of television production was revolutionised with the shift from cathode ray technology to CCD (charge couple devices) for capturing images that led to small, light and more adaptable video cameras. Lightweight cameras – the 16mm camera and later the video camera – make the camera operator's view dynamic, able to move with the head and the body but at the cost of the image being shaky and vertiginous to watch. The invention of the steadicam in 1976 meant that movie cameras became more mobile; attached to the body of the operator but with a counter-balanced movement, they reduced shake and meant the operator and camera could move more naturally. By substituting a video monitor for the viewfinder, the steadicam began a separation of the camera from the operator's eye. The camera on a steadicam, a crane or a skycam produces unusual angles and can produce 'inhuman' movements that are vertiginous and interesting; a view from way below or above normal eye height of a stage performer, an audience or the play on a sports field. Adding computer-guided control to pre-programme patterns of movement, the speed and moves of the crane or skycam can enable it to 'fly' and achieve an exhilarating view of the world not available to the human body.

What John Thornton Caldwell (1995) calls 'televisuality', and Jeremy Butler (2010) 'television style', was a self-conscious attempt to make use of the possibilities of cameras to enhance what the viewer could see, to introduce an aesthetic element to the spectacle. Those who made programmes – commissioning editors, producers and technicians – wanted to make what appeared on the screen as attractive and engaging as possible. Televisuality, as described by Caldwell, was a coming-together of material practices and institutional pressures that was indicative of a broader period of transition in the mass media and American culture (1995, p. 5). Making television, especially cable television, was a quick and cheap way of generating income through advertising as against the high capital costs of making movies with their falling returns. This led to the investment of money, energy, skill and technology in television as an industry, but to attract loyal viewers – and so the investment of advertisers – TV companies needed a distinctive 'look' for their programmes. Designing the right look for a programme during the 1980s could lead to a loyal niche audience who would acquire the cultural capital to enjoy and appreciate its style.

Technique

The techniques of televisuality were initially linked to the use of analogue video but became greatly enhanced with digital recording and editing. The political economy of the media has long since moved into a different phase in which what is produced both for cinema and for the small screen has been changed by the televisual aesthetic (Caldwell, 2008). However, the excess of televisuality has survived into the second decade of the 21st century, marking out a particular aesthetic characteristic of some television programmes. The use of slow-motion graphics in a series like *CSI: Crime Scene Investigation* (CBS, 2000–present) that, for example, apparently shows the trajectory of a bullet through a body, is part of the programme's identity and a display of how television can reveal more of the world than we could see if we were actually there. Televisual style can show material cause-and-effect processes such as the internal workings of a body. However, as film studios have incorporated digital technology, the distinction between movie and television production has become less based on the equipment and techniques and has more to do with the stylistic form of the medium and its practitioners. There are a number of techniques that are not exclusive to television or to film but which have become characteristic of the moving image. Early television cameras had a cathode ray tube, with three lenses for distance, middle and close-up shots that set a limited style. The zoom lenses on modern television and film cameras enable rapid and smooth movement from distance to a view so close that the naked human eye could not focus. The human eye can alter focus for the centre of vision in a way that is unnoticeable, but a contemporary camera technique of a visible change of focus between close and far objects shifts the screen viewer's attention perhaps from one speaker to another when they are in line, or from a near object to its setting. An area of an image that is beyond the focus field and so blurred can be used for aesthetic effect, with colours, movement and shapes juxtaposed with whatever is happening on the sound track or a visual area in focus.

Although the production of content for the small screen may require just one person with a phone camera, a complex team of specialist professionals contrives most of what is watched. They usually work with a script and plan the preparation, shooting and editing but may then modify and bring various elements together as the programme is made. For example, the feed from a number of cameras is blended by a director for a game show in a studio or a comedy programme recorded in front of an audience in a way that gives the effect of a live performance. Since the 1980s, the techniques for stored digital video 'post-production' effects (Caldwell, 1995, p. 138–139) have developed so they can now be done live. Sports programmes blend different sorts of material – graphical information, live feed, action replay, expert commentary – so that interspersed in a live programme is carefully edited content, generated during the broadcast.

Screens

The development of viewing screens was also important in creating the televisual aesthetic of the 21st century. Whereas the early cathode ray, black-and-white screens with rounded corners are iconic symbols of low-definition television, a large, high-definition, flat screen can be as effective in mediating reality as a clear window; what is on the other side of the glass could, for perceptual purposes, be equally real.[3] The idea of representation as equivalent to looking through a window dates back at least to Alberti's 15th century theory of painting[4] but has more recently been coined in relation to electronic screens by Lev Manovich (2001). The 'viewing regime' of the window/screen has a frame that excludes all that is not within it: 'It exists in the normal space of our body, and acts as a window onto another space. This other space, the space of representation, typically has a scale different from the scale of our normal space' (Manovich, 2001, p. 95). The scale of the represented image within the 'window' can vary as the camera captures the 'other' space in a long shot, a middle shot or a close-up. The effect of the screen, especially the small screen of the television, has a fixed relation with the viewer's body so that she is dependent on the context of the frame to judge the size and relative distance of what appears within it. The conventions of realistic depiction within the frame were long ago established in painting and provide a way of engaging with the small screen image. Pictorial representation, whether by painting or photography, does not precisely reproduce the conditions of the eye perceiving the world (it is not stereoscopic, for example), but both involve distortions that human perception adjusts for.

Merleau-Ponty (1962, p. 260) explains that human perception is a product of our having a body through which we orient to size and distance as our intentionality and interest organise our focus of attention. Because we use the same perceptual resources to see the people, objects and scenes on a screen that we use for the world around us, we see them as 'real', but different ways of lighting, colour or shading can affect what we see. For example, Merleau-Ponty describes how the framing effect of a screen or window that obscures the source of light can change the colour of an object so that it appears as a patch of colour – in his example, a feebly lighted white wall becomes bluish grey (1962, p. 307). He is pointing out that colour is not an intrinsic quality of an object but is perceived as reflected light from an object in a context. The effect of the frame changes the nature of perception, and the lighting of a scene changes what viewers perceive (see, for example, Butler, 2010, pp. 41, 71). Like realistic drawing, painting, cinema and photography, television can be a vehicle for visual illusions, but if they are to work, it must continue to invoke a sense of the real that is being distorted. As Merleau-Ponty puts it, 'If myths, dreams and illusions are to be possible, the apparent and the real must remain ambiguous in the subject as in the object' (1996, p. 294). In other words, we have to engage with what we see as if it is real, even if we are not sure that it is. This may be a dangerous way to drive a car, but is entirely appropriate for such a safe activity as watching television on a small screen.

The viewing regime of the small screen demands a suspending of disbelief that accepts the reality portrayed through it, and '...the viewer is expected to concentrate completely on what she sees in this window, focusing her attention on the representation and the disregarding the physical space outside' (Manovich, 2001, p. 96). The suspension of disbelief means that we take seriously what we see happening within mimetic imagery on the screen, if only for the period while we are watching. Jean Baudrillard saw that as the distance between reality and the reality-on-the screen is reduced, there is insufficient distance to make moral judgements: 'There is no separation any longer, no empty space, no absence: you enter the screen and the visual image unhindered. You enter your life as you would walk on to a screen. You slip on your own life like a data suit' (2002, p. 177). Baudrillard's critique is a warning, but moral judgements remain possible while we can distinguish between reality through the screen and the paramount reality of lived experience (see Chapter 5, pp. 107–8, below).

If we imagine a test of the technology in which there are two glass panels, each at the end of a two-metre rectangular tunnel of the same dimensions,[5] one a window directly onto a scene beyond, the other a screen showing what an ultra-high-definition digital video camera was seeing beyond it, would an average human be able to tell the difference? The limitations of the screen technology in delivering a flicker-free, true colour, representation still mean that if the light were good, most viewers could tell which were the window and which the television screen, especially if there were animal movement beyond the screen.[6] But the continually increasing technical quality of contemporary equipment means that this test will soon be equivocal, far sooner than a computer can pass the Turing test. And the television screen will be able to show digitally stored moving video images that are just as easy to mistake for real as real-time images. Already, the capacity for computer-generated images to approximate to 'real' images has moved video gaming away from cartoon towards photorealistic representation, while computer graphics within movies and programmes have been merged almost seamlessly with camera-generated images.[7] Just as we use our moral sensibility to make sense of the real world through a window, so we use the same moral sensibility to make sense of the realistic world through a screen.

Style

Jeremy Butler (2010) has a rather different take on televisual style that is less to do with technology and the political economy of the relationship between television and cinema and more to do with the aesthetic effect of the visual construction of moving images and the cultural capital of viewers in engaging with it. He pays close attention to the average shot length, lighting, camera angles and focus field when, like Caldwell and other commentators, he takes the US police show, *Miami Vice* (NBC, 1984–1989) as indicative of a change in how television programmes were made. Butler shows how the aesthetics of the programme reflected a similar shift to the one achieved 40 years earlier in the film industry by film-noir, and he identifies similar themes of ambiguity of identity and moral purpose; men pass as women, the cops as crooks and the crooks as good family people. What Butler nicely does is to show how, by creating visual ambiguity through televisual techniques, the show creates a moral ambiguity in the reality it portrays. Officials are corrupt, crooks are cops undercover, and the identity switches mean there 'is no clear demarcation between forces of good and those of evil – or at least the distinction is constantly changing' (Butler, 2010, p. 75).

Film-noir often uses dark, night-time settings with deep shadows and high-contrast lighting, but in *Miami Vice,* the technique is sometimes reversed; although shot in colour, high contrast between black and white is emphasised through the brilliant white of sunlight, white buildings, white clothes and pastel interior decor. In night scenes, lighting picks out key characters and objects against a dark background. Exaggerated high or low camera angles, out-of-focus objects in the front of scenes that emphasise action deep in the visual field and self-conscious use of the framing effect of the camera and of objects within the scene are other techniques that Butler describes. His analysis distinguishes the 'attenuated continuity' of traditional narrative television, including soap operas such as *Dallas* (CBS, 1979–1991), and the aesthetically more interesting, but more difficult to watch, 'intensified continuity' of film-noir and post-1980s television series such as *Miami Vice* (Butler, 2010, pp. 83–86). This same intensified continuity is also characteristic of advertisements that are, in effect, intensive mini-programmes in which a high volume of skill and resources are invested to both engage and distract the audience long enough for the message to get through (see Butler, 2010, pp. 109–37).

That television has its own aesthetic is important for two reasons. The first is that it appeals to the viewer, keeping her attention, stimulating pleasure and interest, offering something that either resonates with what has gone before or creates something new and attractive. Appealing to the viewer means entertaining and amusing her, giving her direct and immediate visceral pleasure – which might involve shocking or horrifying her in the safety of her own home. The second reason for thinking about the aesthetic of televisuality is to recognise that it involves a varying relation with reality. What is created within televisual aesthetics is a way of seeing the world that is different from how the world about us is perceived; televisuality opens up a different realm in which reality is other than we experience it. This does not mean that the televisual is wholly fantastic, because it must always retain a grounding in reality as we perceive it if it is to work as communication. Nonetheless, televisuality is non-discursive and non-literary and is about the creation and communication of images, which may include visual ideas, rather than the articulacy of text, whether written or spoken. In this sense, televisuality is 'non-intellectual'; it does not give reasons or rational communication or expression of arguments but does stimulate the viewer's imagination, taking her to another place, putting her in the position of a person or a character, who might be an animal or even an imaginary monster.

Televisuality is the aesthetic dimension of television, the creation of images with sensori-emotional values that move the viewer. The aesthetic dimension can be contrasted with the poetic dimension, the construction of narrative form with coherent meanings that together complete the arc of a story. As televisuality has advanced the possibilities of stimulating the viewer's sensori-emotive sensibilities, then the pressure to achieve a satisfying poetic form has been reduced. The form of television – the in-the-moment of liveness, the continual return of repeats, the postponement of completion with the serial, series and seasons – generates a loose and open poetics and it is able to do this because the audience are so satisfied with the aesthetic pleasures that they make up for the lack of closure, of final endings. Early television was more likely to offer a discreet programme, a play or documentary, that told its story within the allotted period of half an hour or an hour. But as televisuality has opened up the range of televisual communication, there is less need to complete a story; the ending is always a cliffhanger of some sort, an invocation to 'watch this space'. If televisuality does not itself have any moral importance, what it does is extend the possibility of ambiguity in moral messages within a programme. Not all characters need to get their just desserts, not all story lines need to be completed. Information, ideas and impressions can be left ambiguous; sometimes tantalising, sometimes irrelevant. This enables television programme makers to resist the simple or direct moral messages of the completed stories characteristic of literature and cinema.

Spooks

Since 2002, the high-action thriller series *Spooks* has used many of the aesthetic features of televisuality – even though it is shot on film – to show dramatic stories about how MI5 might respond to the threat of international terrorism in the post-9/11 world. The highly stylised 'show' has carefully designed, constructed and lit sets and actors who look good and are well groomed and dressed fashionably. The lighting emphasises contrasts between interiors, especially the headquarters, known as 'the grid', and the tastefully chosen outdoor settings. Each episode follows at least one threat to UK or international security posed by a terrorist organisation or a 'rogue' state. One recurring theme is whether individuals have given up moral responsibility for their own actions and agreed to act on behalf of the state or on behalf of the team and its leader. Narrative tension often revolves around questions of trust between members of the team and their political masters, but storylines feature the exceptional skills, courage and coolness under stress of its characters and the technological gadgetry they use. The gathering and processing of surveillance on those who constitute a threat is a key element in every operation and leads to distinctive imagery, particularly of the display of documents, data, mugshots and maps as well as websites and television programmes on multiple screens.

Explosions and car chases are the stock in trade of many action thrillers, but in *Spooks* they are interspersed with following the action through surveillance cameras, audio 'comms' links, dynamic maps and online building plans. Much of the interaction between characters is through these devices, and much of the 'information' is emphasised visually; the MI5 team even find some things out by watching *BBC 24 News* on a screen in their base. Television becomes a character in the television show. The glass of the computer screen lit by information or images is used to reflect characters' expressions and reactions, while windows on cars and buildings, glass table tops and panels on 'the grid' allow seeing through from one space to another while at the same time getting a reflective perspective on the space the viewing is from. The compound, overlaid images created by reflections are reminiscent of cinematic montage and add moral ambiguity; rather than just taking a character's perspective, we also have a perspective on the character. In one memorable sequence, those on the grid watch a 'satellite infra-red' plan-view image of people moving in a darkened building which the viewers can see but the field agents cannot (*Spooks*, BBC One, Series 5, Episode 8). At times, there are two or three separated images on the television screen showing us agents in the field, back on the grid and in the infra-red plan view. In a later sequence, we see Adam Carter, a key character, paralysed and apparently having a breakdown, intercut with a rapid series of 'flashback' clips representing a mixture of his memories and fears. The complexity of an out-of-control experience of the unconscious bursting through into consciousness is represented in a contiguous montage that makes serial connections both within the programme and to earlier programmes in the series. Under cover of this televisually aesthetic sophistication, *Spooks* is able to raise difficult issues about the state and its use of force, the wielding of political power through covert agencies, managing the threat of terrorism and the negotiation of political interests on an international stage in which violence is always an ultimate threat.

Flow

The aesthetics of televisuality enable individual programmes to raise moral issues without resolving them, leaving the viewers with an ambivalence that would be unsettling if the pleasure of the imagery were not so satisfying. Another way in which moral complexity is sustained by television is the contrast between themes and issues as they change from one programme to the next in the broadcast flow (Williams, 1992; Dienst, 1994, pp. 3–35). Raymond Williams (1992) contrasted the temporal flow of television content with print media, such as newspapers and magazines, where the reader can scan articles and pages to choose what they will give their attention to. Headlines and pictures as well as division into sections (for example, sport, finance, international news) guide the reader in choosing what to read and when. In contrast, the television viewer joins the continuous planned flow of broadcast content in which one programme follows another with mini-programmes in between. As television has evolved, the viewer's choice has increased (different channels, time-shift sources, recordings) so that they can compile their own flow – but it is always derived from the broadcasters' planned flow. It would be possible to watch two screens with different programmes simultaneously, but the viewer's perceptual senses can only satisfactorily absorb moving audiovisual imagery from one source at a time. (When a second screen is used, it is usually for text based communication.)[8]

Despite the proliferating options for personal choice by viewers, there is no sign that the broadcast 'flow' is going to end. Rather than reducing the amount of television watched, it had increased in the US by 22 mins per month in 2011 as against 2010. Whereas users watched on average 158 hours of broadcast television per month, they only spent 10.75 hours watching time-shifted television and 4.5 hours watching video on the Internet (Nielsen, 2011, p. 5). Television companies plan and construct the sequence of programmes that they broadcast, and their broadcasting policy affects what is available on other platforms (repeats, time-shift transmissions, DVD releases). The schedules of what is to be broadcast continue to be published in daily newspapers, online and in specialist publications, and these media preview and review programmes before and after broadcast. The 'flow' of broadcast television continues to be oriented to the daily lives of viewers (daytime, children's, peak time evening and late-night viewing) with a mix of genres to maintain variety with regular punctuations of news and weather information. The major broadcasters compete for viewers with new material and carefully planned repeats to capture and keep viewers for a 'session' rather than a single programme.

As we saw in the viewing session 'An evening in' in Chapter 1, the broadcast flow is made up of programmes of different genres that fit into scheduled slots, usually of an hour or half an hour with sufficient space in between for what I have called 'mini-programmes' (for example, adverts, idents, announcements). The continuous sequential flow is then made up of differently-sized units that would include frame, shot, clip, scene, segment, section, programme and schedule. 'Shots' are continuous sequences of moving images captured by the camera – whether film or video, analogue or digital – but the larger units are usually made up of smaller units edited together through the intentional action of programme makers. Programmes can be categorised according to convention into genres including: news and public affairs, features and documentaries, education, arts and music, children's programmes, drama, movies, general entertainment (variety, quiz, chat, games, comedy), sport, religion, publicity, commercials (Williams, 1992, pp. 73–5), and we could add 'reality', food and nature as genres that are more significant now than in the 1970s. However, Williams (1992, p. 80) argued powerfully that genre was an inadequate criterion of the quality of a broadcaster and neither is it a determinant of the moral importance of a programme; a comedy or cartoon show can deal with behaviour and action of moral significance as great as that of a news feature or a religious broadcast. It is 'planned flow' that for Williams is 'the defining characteristic of broadcasting, simultaneously as a technology and as a cultural form' (1992, p. 80).[9] Cultural analysts of television programmes who have often focused their attention on one of the genre categories, whether it is sport, drama, soap opera, documentaries or reality television, have tended to overlook this injunction. Even as the cultural form of television converges with other media, particularly with audiovisual content on the Internet, and viewers have increased control over when and what they watch, the planned flow of broadcast continues to shape the aesthetic and stylistic form of television (Kackman et al., 2011).

Contiguity

The programmes in a planned flow that abut or are contiguous with each other interact with each other even when their content or genre is seemingly unconnected. This, as Williams noted (1992, p. 87), is because broadcasters who want viewers to stay tuned to the flow of their channel *plan* the flow. For example, an advertisement or a trailer might not be too closely related to the following main programme but have a topic or narrative form that can be connected, probably subconsciously, by viewers. Like Williams, I want to emphasise the sequential flow of television, but as he recognised, it is very difficult to say much in any detail about the flow – and what can be said 'is unfinished and tentative' (1992, p. 90). It is much easier to talk, as I have done in Chapters 2 and 3, about the content of specific programmes or even the sections of programmes, abstracting them from the flow and treating them as typical of a category or genre. As Williams reminds us, viewers get caught up in the flow and are reluctant to leave just because the programme they began to watch ends. One of the consequences is that viewers of broadcast television are exposed to different types of content than they intended – they join the flow, which has been planned to engage them, and they remain with it, even though the moral content of subsequent programmes is quite different.

The 'Evening in' viewing session in Chapter 1 is of contiguous programme and mini-programmes that are each discrete and separate in their content but are immediately adjacent to each other in the temporal flow of the broadcast. There is no gap, no interval; during transmission, the flow is continuous and without break. This flow of temporally contiguous elements is reflected in the structure of the programmes themselves that are made up of sections that are contiguous and have a thematic link but have different topics or narrative threads – this structural form was true of all the programmes in the 'Evening in' sequence, though it is not necessarily true for all programmes. Each programme may have a moral theme which coheres the different 'items', 'cases', 'stories' and 'sub-plots' that make up the contiguous elements. On *EastEnders*, there was a theme of trust amongst family, friends and workplace; on *Panorama,* it was the risk of trying to get a 'dream home' in an uncertain economic climate; on *Crimewatch,* it was the injunction that those who know something about a crime should tell the police. Contiguity brings with it not only juxtaposition of the genre and format of programmes in the flow but also of the moral issues and themes. The different programmes in the flow of a viewing session all contribute differently to the moral imaginary. Even if viewers are channel surfing or blending previously recorded programmes, the effect is a mélange of moral themes and issues; the compound sequences of contiguous moving audiovisual images may make up a continuous flow, but the mores and morality that can be derived are not. Unlike the sermon, the lecture or the feature film, or even a single television programme, the broadcaster takes no responsibility for coherence of a viewing session; there is no introduction or conclusion, no overview or summing up, because viewers begin and leave the flow for very different reasons at different times.

Seriality

If contiguity is the abutting of unlike items in the flow of television, seriality is the connection between items that are not adjacent in the temporal flow, as when the episodes of a serial are broadcast at the same time each week for a number of weeks. Seriality was a feature of 19th-century fiction in magazines by writers such as Charles Dickens, Willkie Collins and Arthur Conan Doyle and early-20th-century children's comics. The instalments of a serial are traditionally of the same length and have some measure of narrative closure within the episode. But it is also characteristic of the serial that a hook or cliffhanger is used to ensure that readers or viewers seek out the next episode. All of the programmes in the 'Evening in' are series (*The One Show, New Blues, EastEnders, Panorama* and *Crimewatch*), but only *EastEnders* has a clear narrative continuation. Some programmes are serially available on following nights – soap operas like *EastEnders* often follow on for a number of weekday nights, for example. Episodes of a series are either discreet programmes – for example, each episode of *Panorama* is on a different topic, often with a different team making it, but has a similar concern with current affairs – or is a continuation with a narrative, as in the dramatisation of a novel such as *Small Island* (BBC One, 2009), which had two episodes. The seriality of many television programmes is a blend of both these forms as there are narrative threads running between episodes, but each is a complete story in its own right. In this type of show, the characters and the settings become familiar, but viewers do not need to watch each programme in the series to enjoy the narrative closure of a story. Seriality is common to much television broadcast by the same channel where viewers become familiar with the faces and styles of newsreaders and weather presenters, the personae of chat show hosts and the settings and styles of magazine programmes.

A series is also sustained by televisual features of continuity – particularly the theme music and the opening titles, which often include a visual ident that can be used in trailers and in advertising material. Theodor Adorno describes the titles, music and opening sequence as creating a 'halo' effect because they forewarn viewers of the sort of programme they are about to watch (1991, p. 145). The lighting of the studio, the colour palette used for design, the type of furnishings and deocration all help shape the atmosphere of a series before any content is included.

The episodic nature of television series is the most apparent form of seriality that emphasises the familiarity and recognisability of programme content, in contrast to contiguity, which emphasises contrast and difference in planned flow. Television has a number of other ways of realising seriality however, through genre, repeats, references and trailers. As viewers recognise a genre, they are able to anticipate the style of programme and will be able to relate one television cop show to another, noting similarities and differences. (For example, *CSI* – CBS 2000 – present, and *Silent Witness* – BBC1 1996 – present, both build stories around police forensic procedures but in very different ways.) The same show is often repeated at another point in the flow, perhaps within the same week or maybe after months or years, but each repeat reinforces the connections between disconnected moments in the flow. Mini-programme repeats within episodes as a catch-up or taster of earlier or later episodes achieve a similar serial link across different points in the flow.

Television has become very conscious of its heritage and will repeat, reprise and remake shows from an earlier era, allowing generations to reminisce. Successful series return for annual seasons, and some programmes survive transformations keeping plot devices, characters or settings. *Doctor Who,* for example, originally ran from 1963–1989 and was re-launched in 2005 with some original features, including the spacecraft TARDIS, and the conceit that the Doctor could transform and the role taken on by a different actor at the beginning of each series. References to earlier shows reward the consistent viewer with a sense of 'insider' knowledge, but even if they did not see the earlier show, they will be made aware of the serial continuity across different points in the flow separated by many months. In *Doctor Who,* some of the alien species, such as the Daleks, are 'brought back' in different series, and in a show like *Crimewatch* (see Chapter 1), there are sections reminding you of previous crimes that have been reported and bringing you up to date with the progress of the investigation. These references between programmes are common in news and current affairs will often include re-broadcasting previously shown material. In *The One Show* (see Chapter 1), there was a reprise of a news story from 20 years earlier, bringing viewers up to date on an inquiry into the cause of a water pollution incident with footage filmed at the time. The cinematic tradition has incorporated temporal devices from other media, such as the flashback, flash-forward and time-jump used in novels and films. But television involves flashbacks and flash-forwards not just *within* a programme but across series, serials, the evening's viewing, the week and the history of the medium.

Conclusions

Although Caldwell's (1985) influential use of the term 'televisuality' identifies a moment in the historical, economic and technological trajectory of the medium, I have taken it to refer to a set of phenomenological characteristics that are to do with how the moving audiovisual image is encountered on the small screen. How television appears on the screen impacts on the way that it contributes to the moral imaginary. In its early days, television could be distinguished from film, not only through the size of the screen and the context of viewing but also through the poor quality of the image. Technological developments have not only improved the definitional quality of television, they have led to the creation of a style that, at least for a period in the late 20th century, was marked by an excess of tricks and techniques that distinguished it from movies. However, continued technological developments, especially digital photography and computer graphics, have led to a convergence of movie and television styles as both have affected each other. 'Televisuality' no longer usefully refers to the style of the image but can still point to the features of audiovisual imagery on the small screen.

Firstly, televisuality is most prominently associated with *live* moving imagery with synchronised sound. This characteristic mode of television from its early days continues today to make sense of news and sport broadcasts and lends a 'realistic' quality to many other genres, including current affairs and studio shows even if they have actually been recorded, maybe even on film. The live introductory sections of programmes like *Panorama* and *The One Show* (see Chapter 1) lend to the filmed and recorded sections a presence and liveness that is very different from a cinema film.[10] The liveness of television makes its moral content relevant to the viewer; it is a view of their society that shows its folkways and mores, its norms, customs and practices. Secondly,

Raymond Williams's concept of 'planned flow' continues to shape the contents of the small screen in a way that distinguishes it from cinema, despite the plurality of platforms and sources of television imagery. Unlike cinema or theatre, where the viewers give themselves up to the event of a specific performance, the television viewers join a flow that continues even after they have left it. As Williams put it in 1974, 'We can be into something else before we've summoned the energy to get out of the chair, and many programmes are made with this situation in mind: the grabbing of attention in the early moments; the reiterated promise of exciting things to come, if we stay' (1992, p. 88). Flow exposes viewers to the moral content of programmes that they might not otherwise have chosen or sought out. Thirdly, contiguity abuts programmes of unalike contents in the flow of a television viewing session, so creating a characteristic diversity while at the same time, the various modes of seriality link items that are dislocated in the temporal flow. The contiguous juxtaposition of unalike elements means that contrasting moral themes and perspectives are presented close to each other, rather than a continuous, coherent moral outlook. The various ways in which serial connections are made across the broadcast flow allow viewers a sense of familiarity and recognition, as issues and possibilities they are interested in return and can be revisited in the light of other experiences. This often means that there is no closure, no final end that reveals, once and for all, the moral consequences of actions; even characters who seemed 'dead' can be resurrected. Whereas narrativity is the meaning structure of the novel, the play and movie, the more or less planned flow of seriality and contiguity is the meaning structure of watching television.

Fourthly, there is a continually developing televisual 'style' that is partly shaped by the technical possibilities available to programme makers, partly linked to the aesthetic capital that is built up in viewers and partly linked to the political economic need to compete for attention with other media and amongst broadcasters. If the flow of contiguous, but different, elements creates the capacity of television to present viewers with a range of moral situations and possibilities in each viewing session, televisuality creates an ambiguity and ambivalence in the outcomes of actions and in the moral worth of characters that opens up different provocations to moral sensibilities and different contributions to the collective moral imaginary.

5
The Phenomenology of Television

From image to feeling

Just how does television communicate with its audience? In this chapter, I will explore the phenomenology of television by which I mean the *process* by which viewers give their attention to moving images on the small screen and make sense of what they see. What most people see in such moving images, especially when there is synchronised sound, makes sense to them in a direct way that affects them, often viscerally. It may make them laugh or cry or feel shock, surprise, anger or other less extreme emotional states such as curiosity or irritation. They may learn things from the television or see things in the world differently than they had before. When television holds viewers' attention, it is usually because the experience is pleasurable, and they are amused and interested, although it may be because they are fascinated by something they find horrific or ghastly. If viewers did not derive pleasure from watching – even what horrifies them – they would not watch; viewing television is something that is seldom, if ever, imposed on anyone, and it is very difficult to imagine how a person could be force-fed the contents of the small screen while they have the power to turn it off. Television is a medium of communication, a device that carries information transmitted in sound and light, but for viewers it often feels *immediate*; once immersed in the process, which requires very little effort or preparation from them, the embodied and material process of watching is often forgotten. Children get the hang of television very quickly without needing to be taught how to watch it, and people who have never seen it before can engage in its delights without introduction or training. If it did not work in this way, people would not watch television with the regularity and the interest that they do; they would not pay attention to the flickering light and the babble of sound emanating from the box with the screen. So how does it communicate? How does it connect with people? I want to argue that television represents the world well enough for people to make sense of it in broadly the same way as they do the ordinary world around them. This is not the way that television has usually been treated by academic commentary that has tended to see it as being a special medium of communication that is potentially rather dangerous precisely because it is so accessible. The danger is largely that viewers might not notice the effects that it is having on them as individuals so it could be changing people in ways that are not good for society. Commentators, analysts and critics have tried to reveal the dangers of television, to investigate how it works and to lay bare the unconscious or unnoticed impact it has on those who watch it.

Effects?

There is a substantial literature that tries to isolate the process by which broadcast material is taken in by the viewer, to isolate and identify its 'effects'. (For overviews, see Cumberbatch et al., 1987; Felson, 1994; Gunter, 1994; Gauntlett, 2004 and Chapter 2 above.) For example, the showing of violence – harm, hurt, injury, mutilation and death – may have a number of possible effects which could be different for different people. Such imagery might lead to the imitation of violent action, it might legitimate an existing impulse or tendency to commit violence or it may lead to a tolerance of violence in other people. On the other hand, it may have the opposite effect of showing viewers the harm caused by violence, stimulating their disapproval and promoting disgust and indignation. These are quite contrary moral reactions to the imagery, and it is unlikely that any programme with violent content will have a specific 'effect', since whatever is viewed is always viewed in the context of other programmes. But more importantly, it is viewed in the light of the viewers' own values that they have acquired from their culture and their experience – including watching television. The response to imagery is also ultimately affected by characteristics of personality that include the inclination to watch a particular programme. Somewhere in this complex set of processes is the issue of aesthetic taste that often feels like a personal choice but is also shaped by cultural milieu – taste varies according to class, race, educational experience and gender, for example, but it also changes with history and over the life-course of the viewer. The difficulty with the notion of 'effects' is the implication of a uniformity of effect on viewers, of a determinate relation that might be predicted, perhaps within certain constraints, and therefore could be judged as 'good' or 'bad'. As I suggested in Chapter 2, while it is reasonable to presume that television – whether specific programmes or the media as a whole – has effects, it is very difficult to be sure what those effects actually are.

As if the complexity of what the viewer brings to their viewing weren't enough to make understanding effects so difficult, there is a further issue of how viewers actually *view*. Different viewers at different times will engage more or less consciously with what they see and be aware of thinking about it, even uttering, speaking or expostulating, in response. At other times, the same or other viewers may be passive, less conscious and apparently disinterested while still glancing or gazing occasionally. Some may be emotionally moved while others remain impassive and apparently unaffected – though this does not necessarily mean that what they have seen has had no effect at all. It is reasonable to argue that what is shown on the screen will stimulate interest and attention or may lead to distraction or inattention, but once again, the viewer and the context of viewing will affect and alter how any programme is engaged with.[1] Nonetheless, the audiovisual content of television communicates with human sensory apparatus in a way that it does not with other animals. Animals may respond to television – especially the noise or movement of other animals – but they do not watch it as humans do; making sense of unfolding events and understanding sequences of action.

If you look at a person who is watching television, you are likely to notice their blank, vacant, unmoving, fixed-focus stare, face bathed in the bluish light emanating from the screen, body relaxed, often slumped in a chair, hardly moving except for occasional twitches and shifts. It is difficult to judge what effect their viewing is having, and it is easy to imagine that the beam of light coming from the screen is projecting through their eyes and directly into their minds. Jerry Mander (1978) does not spend much of his book discussing or analyzing either the content of television or the institutional and organisational processes of the medium, but he argues that television offers a sensually limited experience of the world that colonizes and restricts the viewer's mind. The emanations of light from the television screen form images that he asserts are damaging to the human body and are perceived by the mind in a 'kind of wakeful dreaming': 'Television inhibits your ability to think but it does not lead to freedom of mind, relaxation or renewal. It leads to a more exhausted mind' (Mander, 1978, p. 214). Mander argues – as others have – that the medium of television, unlike reading a book, does not engage thoughtful and critical responses and cancels or counteracts viewers' own personalities because it uses the moving image (see also Postman, 1985). If he is right, of course, television would be a very dangerous medium with the potential for mind control on a mass scale, but his argument is too crude and reductive to be taken seriously; those who watch television are not a breed of uncritical people incapable of thought, lacking in personality and left like zombies by their exposure to the small screen. Are you?

Influence and propaganda

Watching moving images on small screens with coordinated sound is a powerful tool of communication, but there are no grounds for presuming that it necessarily leads to domination of the viewers' minds. There is a risk, however, as with all media, that if the content is artificially restricted (for example, through rigid state control), then it could have a simple propaganda effect of promoting one ideology. Even then, of course, some will disagree, reacting to the content and seeking ways to express an alternative view. The social cultivation thesis of George Gerbner and his colleagues (Gerbner and Gross, 1976; Gerbner et al., 1994 – see above, p. 22) about mass media modifies the simple cause-and-effect model of laboratory studies on effects, but it still sees mediated communication as one amongst a number of distinguishable causes. A different approach comes from research on the impact of mass media on voting behaviour that looks at 'influence' rather than cause. Researchers in the 1940s and 1950s recognised that if the mass media could affect voting behaviour, then this would give them enormous power to shape the political arena and threaten the principle of democracy in which the people share power equally. Again, rather than finding clear and unequivocal effects that suggested that the media – in this case, newspapers – were able to determine the outcome of an election, the finding was more ambiguous. One of the most famous of these studies came to the conclusion that '...the most partisan people protect themselves from the disturbing experience presented by opposition arguments by paying little attention to them. Instead they turn to that propaganda which reaffirms the validity and wisdom of their original decision – which is reinforced' (Lazarsfeld, Berelson and Gaudet, 1944, quoted in Klapper, 1960, p. 20). As well as reinforcing existing ideas, however, the media seemed to have an indirect effect by being mediated in turn by local opinion leaders, people who already had the respect of a social group at work, in the family or a circle of friends or associates (Katz and Lazarsfeld, 1960). Opinion leaders were shown to be important in interpreting and evaluating ideas and then passing them on through face-to-face interaction. In summarising this media effect, Klapper (1960, p. 20) points out that to be such a leader, the person must be seen to reflect the views generally shared by the group and be a personification of certain values.

Another way of thinking of 'influence' is that by working with those who have power and the capacity to influence others, television and other media can work as propaganda. This is precisely what Edward Bernays called the 'new propaganda', which he described in terms of the artful persuasion of designers, society people, editors and critics to influence fashion amongst consumers: 'Only through the active energy of the intelligent few can the public at large become aware of and act upon new ideas. Small groups of persons can, and do, make the rest of us think what they please about a given subject' (2005, p. 57). His idea of propaganda was based on advertisements of all sorts, spectacles and events and the exploitation of notable people and personalities to promote a 'product', whether it was a car, a politician, votes for women, art or education. And the new propaganda meant that business, politics, activism, aesthetics and academia could all learn from each other which promotion techniques were effective. The mechanics of propaganda, as Bernays emphasised, are based in the communication media, especially the mass media of newspapers, magazines, radio and the movies that those in public relations can influence.[2]

In fact, in democratic capitalist societies, the media's interest is in having as many readers, listeners and viewers as possible, which conflicts with propaganda that tries to influence the audience in a particular direction and so will put some off. The media want to maximise their potential number of consumers with a variety of views and ideas – unless, of course, they are owned and controlled by the state. Williams (1974, p. 126) warns us that to criticise television for exposing the public to a dominant set of normative meanings and values as propaganda, overlooks the role of priests, judges and teachers whose personal influence is promoted as a good way in which social values are passed on. That television is presumed to have *some* influence is demonstrated not only by commercial advertisements and sponsorship but also by political parties investing in broadcasts and manoeuvring to maximise media time for their spokespeople. The belief in the impact of television is also entailed in broadcasting policy, such as the UK's 9pm 'watershed' for certain types of programme content, that treats children as less resistant to media influence than adults. In 2010, a minor 'moral panic' (see Chapter 7, pp. 153–6) was set in motion by some revealing costumes and sexually-suggestive dancing by the pop stars Christina Aguilera and Rhianna on *The X Factor* (ITV1), a British talent show on family television broadcast before the watershed. Ofcom, the broadcast media regulator in the UK, received 2,868 complaints about the sexual content but deemed 2,000 of these to have arrived after the *Daily Mail* had published pictures and claimed that there was viewer outrage about the performances. Ofcom decided that the broadcast had not breached guidelines but that elements of the performances were 'at the limit of acceptability for transmission before the 9pm watershed.'³ This example shows the concern of viewers about the influence of television but also the influence of a newspaper in stimulating responses to a television programme.

'Reading' television

Anthony Smith distinguishes between those approaches to the effects of television that focus on the individual – the gratifications they received from viewing or the influence it had on them – and those that focus on the society – the setting of an agenda of issues and the creation of meanings (1985, pp. 11–12). If it is difficult to identify the effects on individuals of watching the small screen, perhaps there is more mileage in trying to make sense of it as a creator of symbolic meanings that are common to everyone. This approach follows the well-established field of literary criticism in which the possible meanings of a text are interrogated and interpreted; even if not all viewers receive the same messages from their watching, the text can be treated as a stable, coherent and consistent whole in which any viewer could find the same meaning. In a classic study of a single programme, Stephen Heath and Helen Skirrow (1977) broke the flow into a series of shots that they organised and described, with dialogue and commentary transcribed. Their analysis unravels how the programme was put together, to point out that despite it being a current affairs documentary, its form was fundamentally novelistic; they were challenging the 'realism' of the imagery and emphasising the constructedness of the programme. The possibility of 'reading television' was coined by John Fiske and John Hartley (1978) in their book with that title but has continued to have a serious following (for example, Hodge and Tripp, 1986; Bignell, 1997; Iedema, 2001). The idea of 'reading' television borrows the structuralist approach to cultural products that treats them as constructed out of 'signs' organised into 'codes' of which language is the archetypal instance. Messages encoded in signs are amenable to 'semiological' analysis that will attend to the 'syntagmatic' and 'paradigmatic' relations through which they convey meaning (see Tudor, 1999). Fiske and Hartley draw on the jargon of Saussurian linguistics that gave rise to semiology, and treat television as a 'bardic' form that through 'a series of consciously structured messages' orally communicates 'a confirming, reinforcing version of themselves' to members of the culture (1978, p. 86). They apply a series of concepts including 'code', 'anamnesis', 'function' and 'mode' to the content of television programmes and incorporate Parkin's famous tripartite distinction between 'dominant', 'subordinate' and 'radical' meaning systems of audience reception (see also Morley, 1992).

The difficulty with this approach to television is that it treats its content as a linguistic text rather than as moving images with coordinated sound. As Joshua Meyrowtiz argues, television creates situations and settings in which interaction between characters takes place such that 'to understand televised sequences is far simpler than learning arbitrary codes of reading and writing' (1985, p. 76). The rather tortuous idea of 'reading' television presumes that the continuous flow of audiovisual content can be broken into discrete or discontinuous units to be treated as signs so that their relationship can be analysed (Manovich, 2001, p. 29). It presumes that television is 'encoded', but even though the makers of films and television programmes may include signs of all sorts – including textual information – what is characteristic of their cultural production is the continuous flow of audiovisual representation. Even though that flow is broken into fragments and edited together, its communicative quality comes from its fullness, or what Ellis calls its 'superabundance of details' (2002, p. 12), the photorealistic and audio authenticity that gives it its communicative power:

> There is always more detail than is needed by the narrative; always more present in the image than is picked out by the commentary; always more to be heard than the foregrounded sounds. We see the details of clothes and places, hear the distinction and personal timbre of voices…. we can read the superabundance of details in any image or sound as the proof of its authenticity or the indicator of atmosphere. (Ellis, 2002, p. 12)

What the small screen presents us with is a level of detail and complexity that is too great to be decoded like the music written on a stave or the typewritten script with direction notes, and the volume of detail arrives in our senses of hearing and sight, simultaneously, giving us the opportunity to feel as if we were there. Alfred Schutz (1971, p. 324) draws – as

Meyrowitz (1985, p. 95) did rather later – on Suzanne Langer (1957) to explain that pictorial presentation is non-discursive because it cannot be defined in terms of other independent or arbitrary signs but conceptually operates as a 'flux of sensations' that are related to what is being depicted by similarity. Without the other senses of smell and touch, the presence of what emanates from the small screen is not as complete as immersion in reality, and it requires the viewer giving themselves up to the medium through watching and listening, engaging with the flow from which meaning is extracted not through a learnt code but through the same perceptual apparatus with which the flow of the life-world is grasped.

Language, both written and spoken, has to be learnt, as do other sign systems, including the traffic light code beloved of semiologists, but television presents us with images and sound so recognisable from the world we live in that we do not need to learn how to make sense of it – it has no discrete code of its own that we must first learn. To understand television, human viewers do not need to begin by breaking it into signifying units as we do when we learn written language. We do not even need to learn it as babble, a familiar pattern with only vague meanings and great ambiguity, as children do when they first learn to speak. At some point, the flickering light attracts enough attention, and what is seen and heard is recognised enough for interest to be engaged. When children learn to speak, their imitative babble rapidly becomes reconfigured as units, words that name things or express relationships, confirmed by ostensive and other gestures. Most people learn the spoken language of their community with little effort or thought, although it is usually in a context of encouragement and instruction from parents and others who will correct errors and help with expanding vocabulary.

In fact, television has come to be recognised as a potential tool in assisting the learning of language; it is not a separate world that has to be introduced to children. There is apparently a 'video deficit' in that toddlers between 15 months and 30 months do not learn as well how to imitate what they see on two-dimensional television images as they do from live, three-dimensional demonstrations (Zack et al., 2009, Barr, 2010). But this does not mean that they cannot make any sense of what they see, and, just as in the world around them, what they make sense of is greatly affected by the intermediation of siblings, parents and other people. When it comes to written language, the learning is much more effortful and is usually structured through a method employed by skilled teachers at a school or nursery. Learning to read written language is demanding and takes time. But learning to make sense of television does not take anything like the effort or instruction; as soon as a small child recognises the content of the two-dimensional images on the screen as faces, people, animals, characters and settings in which the characters interact, then they can become engrossed and derive meaning from it. Small children may show little interest in adult programmes, but they will respond to those programmes whose content reproduces characters, stories and objects that are recognisable in terms of what happens in their lives.

Henri Lefebvre (2002) takes an interesting line on images as he distinguishes them from signals, symbols and signs to avoid the structuralist fetishisation of language as a sign system that undermines the fullness of human communication in everyday life. Words, he argues, involve mediation and do not have an irreducible nucleus of meaning; they are always dependent on the language system of which they are a part. The structuralist focus on language and the arbitrariness of the sign has been at the expense of the various elements of what Lefebvre calls the 'semantic field', the various vehicles of meaning in human culture. 'Signals' are arbitrary but within a non-ambiguous, simple, closed and codified system of meaning that functions for one sense, auditory or visual. 'Signs', such as words, can, on the other hand be equivalent in either the visual or auditory channels, and they are ultimately composed of smaller units (the phoneme, the letter) that are arbitrary and have no intrinsic meaning but combine to produce the tripartite structure of sign, signifier and signified. Combinations of signs (for example, words, phrases, sentences) that become carriers of meaning nonetheless are highly structured, although the resulting order 'oscillates between platitude and rhetoric, between banality and expressivity' (Lefebvre, 2002, p. 282). It is the non-signifying elements of speech – sound, gesture, mimicry, grimaces – that enable discourse to be expressive while the system of language itself is almost, but not quite, reducible to logic. The image, in contrast, has more in common with the affective and cultural aspects of the symbol than with discursive communication that excels in clarifying and distinguishing. The image has continuity with reality and an emotional dimension not dependent on the abstract level of discourse. He writes of the image:

> Like the symbol, it appeals to affectivity; it is born in and emerges from a level of reality other than that occupied by signs and their connections... it has certain of the powers of the symbol, arousing affective complicities and pacts directly, without using representations as such. It makes itself understood by setting emotion into movement, and by arousing it. (Lefebvre, 2002, p. 287)

The image is able to provide a link between the present and the past because of its connections with individuals' past experiences while it also suggests the future possible events that can be evoked in the imagination. Lefebvre accepts that the image needs signs (the spoken or written words that invariably accompany it) to communicate, but it 'overloads' these words with expressive and symbolic content. This means that we cannot reduce the image, especially the complex, dynamic, moving image, to signs or codes that have denotative and connotative meanings assured by convention. The expressive and symbolic content of synchronised audiovisual imagery is not 'read' but viewed or watched. Imagery cannot be simply translated or decoded into other signs because it *shows* rather than *tells*; the process is one of mimesis rather than poeisis.

The phenomenological mode of television

Now there are 'codes' used in television, not least the linguistic codes of speech, and television can be treated as a 'text' for limited analytical or critical purposes. But there is no persuasive reason to believe that 'decoding signs' is actually how viewers make sense of what they are watching. Instead, I want to develop a phenomenological account of television that draws on the photorealistic quality of the flow of imagery and the synchronised audiorealistic sound of what is seen. Jenny Nelson offers a phenomenological approach to television that is inspired by the French philosopher Maurice Merleau-Ponty and which summarises nicely the position I wish to establish:

> Phenomenology reflects upon television's mode of presence to the person instead of television's relationship to the person. … It is the structure and meaning of the televisual as experienced which is sought, rather than the structure and meaning of the televisual in and of itself … Television is taken as it presents itself to a really existing person in the natural attitude, rather than to a representative of a particular demographic or semiotic category. (Nelson 1986: 67)

Roland Barthes, writing in 1961 about the potency of the photograph, referred to the 'analogical plenitude' that makes it literally impossible to describe in words what the image contains without distorting it to the extent of changing its structure 'to signify something different to what is shown' (Barthes, 1977, pp. 18–19). The fullness of audiovisual material is even greater and even more difficult to describe or to reduce to a code or the operation of signs. Barthes shows how connotative meaning or encoding can be added to the production of news photographs, and the same happens with television. Such coding (for example, blue flashing lights in the opening titles to connote the urgency of action that will follow – see *Crimewatch* in Chapter 1) can be resisted, ignored or countered. But audiovisual imagery, like the photograph, has an *immediacy* that communicates at a baser human level and must be effective for connotation to have a vehicle. Instead of trying to decode or read television, I want to suggest that a phenomenological approach is needed, one which does not claim to analyse *the* meaning of any particular programmes but leaves intact the flow of dynamic imagery on the small screen.

A phenomenological approach to television is not new and is an aspect of analyses by Tony Fry and his colleagues (1994 – who explore how television works, using a series of Heideggerian concepts), Richard Dienst (1994 – who draws on Heidegger, Marx, Derrida and many others) and Paddy Scannell (1996 – who uses Heidegger to understand the 'dailiness' of television). But rather than to Heidegger's view of technology or temporality, I will look to the work of Husserl and Schutz about how we understand the world around us, to develop my phenomenology of television. Let me begin with the film theorist Christian Metz (1992), who, although he is usually associated with the structuralist idea of there being a 'language' of film, uses a phenomenological approach to understand how the moving image achieves an impression of reality. Whereas Barthes's (1993) analysis of the photograph suggested that it captured the 'here-now' but was always viewed as a representation of the 'there-then', or 'that-has-been', the moving image is located in the present but spatially separated from the viewer – the cinema movie has a sense for its spectator of 'there-it-is' (Metz, 1992, p. 6).

The mechanical representation of what is in front of the camera has similar qualities of shape, proportion, depth and perspective as it does in real life, but the flowing transition of movement, which need be no more than that of the leaves on a tree in the background, offers something to the spectator which, Metz argues, enables the faculty of perception to be more or less able to 'realize (to make real) the object it grasps. ...' (1992, p. 6). Movement frees the object from the surrounding flat surfaces and gives it a corporeality, a substance that allows it to stand out as a figure against the ground. An object or the figure of a person in motion is in a state of becoming, and the spectator is drawn into the unfolding of its potential, the next moment of its existence.

Metz sums up the difference between the ways theatre and cinema represent reality by arguing that theatre is fictional merely by convention, whereas cinema is inherently fictional:

> The cinematographic spectacle, on the other hand, is completely unreal; it takes place in another world…. The space of the diegesis and that of the movie theatre (surrounding the spectator) are incommensurable. Neither includes or influences the other and everything occurs as if an invisible but airtight partition were keeping them totally isolated from each other' (Metz, 1992, p. 10).

Metz goes on to explain how meaning is formed in the cinema through montage, narrative and diegesis that combine to create an impression of a reality, but one that is separate from the reality that is being lived by the spectator.

These representational techniques are all used in television, and not just when it is showing cinematic films, but the phenomenological operation of television is very different because it originates as a *live* representation that can convey moving images of reality in the moment of emergence – albeit at a distance – something that cinematic film can never do. Live television is inherently realistic, properly a medium of 'here-we-are', bringing unfolding events as they happen close to the viewers where they are. Television has developed a distinctive and powerful mode of mimesis that creates what I will call a 'continuous present' for the viewer that is experienced in a similar way to the continuous present of unmediated reality of the world about us.

For example, in showing a player running across a football pitch, the televised representation realistically imitates the form, colour, size, proportions and dynamism of a figure that a spectator would see from a particular place in the stadium. The 'here-we-are' of the small screen reproduces the same temporal flow of what are recognisably the same action and events, as are experienced by those who are actually present. The effect is to make the viewer more or less engaged in the action so that, despite their actual distance from what is going on, they are able to participate in the excitement or emotion of the moment as if they were there. Individual viewers will situate themselves differently in relation to the experience so the person reading a newspaper or magazine while the television is on may not be engaged at all or may come in and out of engagement with the continuous present on the screen. Television may also work as a background – for doing the ironing, doing homework, cuddling or having a row. But while attending to what is happening on the screen, the action is as if it were present for the viewer, they can participate in the 'here-we-are' with others either in the room with them or with those actually at the event on the screen. The viewer watching a football match, for example, sees the pass and the goal in the same time frame as the spectator who is physically present in the stands. If the match is live, then the temporality of the whole event is shared between those actually present and those watching on a small screen, but even if it is recorded and edited down, there is still the illusion of a shared present for the periods of continuous play that are shown.

When it began, all television was 'live', everything was performed in the same temporality in front of the camera as that experienced by the viewer at home in front of the screen, and it was fully continuous during broadcasting. As more and more television broadcasts are of recorded material, the medium has lost some of its theatricality, its presence in the moment, but it retains a key feature of the live form because presenters and others talk straight to the camera and, even when recorded, the talk show or game show can display the spontaneity of being live. Cinematic movies are normally tightly-scripted with a storyboard and shooting schedule so that everything is precisely planned and prepared. But on television, sports events and discussions appear as if they are live, even when they may in fact be recorded, and live, studio presentation is used to topicalise and make relevant other pre-recorded segments by bringing them into the live moment of viewing (see, for example, *Panorama* and *Crimewatch* in the 'Evening in' in Chapter 1).

Television drama, which is now routinely recorded, retains a feel of its live and performed quality despite the technical devices of modern televisuality discussed in Chapter 4. This is because of the 'perceptual realism' that lies behind how we make sense of television images that, as Noël Carroll puts it, 'are processed by the same visual capacities that we use in everyday life' so that it carries the rhetorical force that 'what it depicts is the case' (1998, pp. 138–9). The verisimilitude of what the camera and microphone capture has a naturalisation effect on what appears on the small screen with the risk that viewers simply accept the distorted view of the world that is shown, and their moral horizons are unable to imagine alternative possible realities. Carroll recognises that realism may be developed as a style to persuade audiences in documentary or news programmes and especially in drama. But he argues that the perceptual realism of the television image does not mean that audiences simply believe what they see. At the very least, they bring an understanding of the difference between fiction and actuality, but they also need mental imaginative activity to engage in fantasy and to construct a narrative from the separate arrays of realistic imagery in the various scenes that are edited together to make a programme. Carroll confronts the possibility that through perceptual realism, fantastic illusions or the hypnotic effects of televisual techniques, television as a medium has a deterministic moral impact, but he concludes that viewers require a 'moral imagination' to make sense of what they are viewing:

> Typical programmes – fictions and news programmes – constantly call upon their audiences to make moral judgements. This presupposes the activity of the moral imagination…. That they do follow such programmes indicates that their powers of moral imagination are suitably engaged. (Carroll, 1998, p. 150)

Carroll's remarks astutely point to the necessary engagement with the moral dimensions of television if viewers are watching with any attention. In Chapter 8, I will develop the idea of a 'moral imaginary', the collective and shared stock of imagery, but here I want to pursue the phenomenology of viewers' engagement with the small screen.

A province of meaning

What happens on the small screen is different from 'paramount reality' in either the sense that William James uses the phrase (the world of sensations in which material things are felt by the body and which have a continuity for the mind – 1950, p. 299) or as Alfred Schutz uses it (the intersubjective world of everyday life – 1971, p. 341). The viewer's embodied relationship with a television device happens *within* paramount reality, and yet the content on the screen opens up another world of experience, inserted into the flow of everyday life, creating a new mode of intersubjectivity. Unlike purely auditory media (radio, telephone, recorded music) or static images (photographs, drawings, paintings), what appears on television constitutes a 'sub-universe' (James, 1950, p. 291) or a 'province of meaning' (Schutz, 1971, p. 231) that draws on, but is distinct from and parallels, the 'paramount reality' of everyday life. Purely auditory or purely visual media are more easily integrated into paramount reality and so have a different phenomenological effect (for example, it is easy to listen to the radio while doing many types of work, and it is easy to take in a photo-based advertisement on a hoarding through a glance while walking or driving). Still images are available to be scanned as much as and when the viewer wishes, but the moving audiovisual imagery of television requires continued attention in which paramount reality recedes as the viewer gets caught up in the sub-universe through the screen.

The 'sub-universes' described by James (1950, pp. 291–2) as alternative orders of reality are either idiosyncratic worlds of opinion or madness or the shared abstract worlds that exist as a form of knowledge, whether it is scientific, ideological or supernatural. For James, the world of media (the *Iliad*, *King Lear* or *Pickwick Papers*) are supernatural worlds of fable to be ranked with those of faith (1950, p. 299). But, of course, television, as well as presenting these sorts of mythical worlds, also presents the worlds of knowledge in the form of 'news' and factual programming. For Schutz, who clearly builds his phenomenological account in the light of James's ideas, symbolic representation at once transcends paramount reality but then is always interpreted in relation to everyday life. He suggests that there is a 'shock' of transition from paramount reality into another finite province of meaning, such as occurs with the raising of the curtain in a theatre, having a religious experience, the disinterested contemplation of the scientist or indeed falling asleep and entering the world of dreams (Schutz, 1971, p. 344). For Schutz, there can be inter-subjective socialisation within finite provinces of meaning, and what I wish to argue here is that something akin to intersubjectivity occurs as the viewer engages with other worlds through the small screen.

The shock of transition between provinces of meaning is something that the experienced small screen viewer is consummate at coping with. Television watching involves the transitions from one programme to another, from one narrative line to another and from one time scheme to another. In reverse of Schutz's transition, the television viewer is more likely to feel 'shock' when their viewing is interrupted by paramount reality, when the doorbell rings or when someone comes into the room wanting to discuss something other than what is on the screen. To understand the experience of the small screen as being a shift from 'paramount reality' to another province of meaning is apposite (see Silverstone, 2007, p. 110) because the focus of attention of mind and body is *through* the screen and into the environment beyond, created by moving images and their accompanying contextual and ambient sound. The person becomes absorbed in the province of meaning available through the television screen, giving themselves up to its twists and turns, its temporality, its cultural references. The viewer's body can go into a semi-somnolent reverie with limbs relaxed in a fully supported seating position as quasi-dream material is generated from outside the person. A mode of partial suspended animation, of entrancement, often occurs in which the body is largely inactive beyond stretching or scratching. But even when attention is not wholly absorbed, and the screen is consumed more at a 'glance' (Ellis, 1982 – see Chapter 4, p. 74), the way sense is made of what is seen and heard is through its reproduction of something akin to the experience of everyday life.

Modern televisions are better able to imitate how we see paramount reality, which is, of course, the standard against which improvements in the apparatus are judged. The television image is (for the moment ...) two-dimensional and does not allow the viewer to alter perspective by moving the head or body. But the capacity of the camera to scan whatever is in front of it recreates the scanning capacity of the viewer, even though their head and eyes are directed into a fixed viewing frame. As a number of commentators have pointed out (for example, Sartre, 1991, p. 11; Ellis, 2002, p. 19), photorealistic images show incidentals and minor details that mean they are more than descriptive of reality. Instead of sketching or summarising, they *re-present* it with a fullness that is consonant with the direct perception of what is shown. The 'realism' of the televisual image is enhanced by sound that is linked to the visibility of speech, action or events but most especially by the reproduction of lifelike movement so that what is perceived has the same dynamic and temporal properties as if it were directly perceived. Both the photorealism of moving images and ambient sound (especially in stereo or surround sound) provide a 'superabundance' (Ellis, 2002, p. 12 – see Chapter 5, p. 100) of information beyond that which is needed for a descriptive, narrative or symbolic message but which reinforces the realism of the province of meaning or sub-universe. It is through these means that the small screen creates an alternative flow of perceptible settings, actions, events, objects, animals and people to that of everyday paramount reality. The viewer can be drawn into it as a parallel form of existence, akin to a dream world, in which actions and interactions and their possible and probable social consequences can be explored without there ever being any consequence beyond the realm of mimesis.

A continuous present

John Thompson describes the 'monological' or one-way capacity for television to communicate across time and space as 'mediated quasi-interaction' (1995, pp. 82–100 – see Chapter 6, p. 121 below). In a similar way, television creates 'quasi-intersubjectivity' for its viewers, an intersubjectivity that resembles intersubjectivity in paramount reality because it seems as though the people who are shown are really there; it is *as if* they are present. The photorealistic representation of people combined with a temporality in which the rhythm and pace of their actions happens in a 'continuous present' parallel to the continuous present of the viewer's paramount reality. The viewer need be under no illusion that this is a *quasi*-intersubjectivity, that the people they see and hear on the small screen are not actually present, for them to feel their presence *as if* they were there. Speech and the bodily movement of people, animals and objects in the moving images have the same continuity for those watching on the screen as for those actually present and watching. As the viewer watches, the formation of speech, bodily actions (including those of animals), natural events and the operation of machines happens in a temporal frame that matches the continuous present through which we experience paramount reality. Complex communicative processes such as conversation, eye contact, gestures and demeanour that are usually lost in the fragmentary instant of a photograph all happen in television within a temporality in which the flow of normal bodily movement can be perceived. Embodied communicative content and its effect on others who are co-present are part of what makes a televised interview, for example, make sense to the viewer; we see the reaction on the faces of those who are listening. (Television is often careful to show us these reactions, even if they are recorded out of real time). Whom to believe and how to feel about what is said cannot be detached from the viewer's witnessing of performance; it is as if they were there, directly observing the exchange within their unmediated paramount reality.[4]

The continuous present allows the viewer to grasp the complexity of 'this follows that' without it being reduced to a linear, causal sequence. Sequence is a characteristic of discursive, narrative forms, and in the telling of a story, whether in speech or writing, the descriptive content, temporal pace and rhythm are rigidly controlled by the storyteller. But with television (including made-for-cinema films), the temporal pace and the aural and visual fullness must be convincing as equivalent to paramount reality, at least for the period of a scene. The interchange between actors, both their speech and their bodily reactions, how props and other non-humans act, must all be easily recognisable as those that would make sense in paramount reality. This has led to the increasing use of locations and filming away from the studio where theatrical sets risk revealing their fakeness; to put it simply, things must happen 'as they would in real life'. Programme makers very seldom leave a scene to simply unfold in front of a camera for any length of time; the rhythm and pace of paramount reality are too slow for communicative purposes. But the effect of minutes, even of seconds, in which action unfolds at a 'natural' pace, with the image and sound detail equivalent to that available to a bystander seeing through a window, is to engage the viewer *as if* they were a co-present bystander.

The potency of the 'continuous present' of film, television, video, theatre and some parts of video games is that what happens can be made sense of in an habitual and subliminal but intentional way – the same way a person makes sense of events in the everyday life of paramount reality. Viewers are aware that the continuous present of television is articulated within a language that includes, for example, intercutting, flashbacks, jumps, chapters and scene changes. If viewing is to make sense – and it does not always, as the occasional questions of our co-viewers attest – then the segments of continuous present sequences have to be joined together in a way that works as a programme (or a show, or a story, or a movie). The technique of constructing television as meaningful narrative often requires that sequences of continuous present video are discursively linked by introductions, titles, labels, graphic devices such as logos and 'idents', a voiceover commentary, a presenter talking to the camera, an interviewer asking questions or a character in a drama summarising what has happened. The power of the small screen is precisely in presenting something that is as manifest and substantial to the viewer's perception through eyes and ears *as if* it were materially there with them in the same space and place. The viewer need be under no illusions about where they actually are to become deeply involved in the mimetic province of meaning through the screen. In fact, the predominant mode of television is based on presenting the experience of 'being there' and in simultaneously commenting on it through editing, camera techniques and commentary and interpolation through voice and graphic text. Lev Manovich coined the term 'metarealism' to refer to the 'oscillation between illusion and its destruction' that is involved in this co-experience of the reality of what is represented and the communicative devices that interfere, interpret and modify it as a modulated form of reality (2001, p. 209).

Appresentation

The continuous present of superabundant mimetic audio/video presents viewers with a mode of reality that they can make sense of using the resources they have acquired for everyday life. Schutz explains how the symbolic system of a culture is passed on as a '...set of systems of relevant typifications, of typical solutions for typical practical and theoretical problems, of typical precepts for typical behaviour, including the pertinent system of appresentational references' (Schutz, 1971, p. 348). The concept of 'appresentation' comes from Husserl and refers to the way that symbols form an association between a representation and the thing itself. Husserl wanted to understand how we interact with other subjects to produce intersubjectivity; other people can be intentional objects for me, but I am unable to directly perceive or experience the contents of their consciousness in the way that I directly perceive my own. However, this does not make the other person's mind totally inaccessible because, at the very least, I am aware that they have a mind and a consciousness and can recognise the other as a person through their behaviour (Bell, 1990, p. 220). Husserl puts it like this:

> There must be a certain mediacy of intentionality here, ... making present to consciousness something that is 'there too', but which nevertheless is not itself there and can never become an 'itself-there'. We have here, accordingly, a kind of making 'co-present', a kind of 'appresentation'. (Husserl, 1960, p. 109 emphasis in the original)

There is an indirect 'making co-present' via the mediation of what the other person is experiencing; their behaviour can mediate their experience to me if there is some equivalence with my own experience. According to Husserl, appresentation is a matter of my perception or *ap*perception and is not a mental or cognitive process, such as an act of thinking or reasoning based on induction or deduction or inference through analogy (Bell, 1990, p. 221). Even though it is not the same as experiencing what is directly present through perception, the person experiencing appresentation does not need to be consciously aware of it. What Husserl calls the *'analogizing transfer'* of apperception is based on recognising the other who is *'livingly present'* as an animate organism who experiences the world in a similar way to myself (1960, pp. 110–1 – emphasis in the original).

Husserl links appresentation to the higher psychic sphere of 'empathy' as, for example, with: '...the outward conduct of someone who is angry or cheerful, which I easily understand from my own conduct under similar circumstances' (1960, p. 120). Appresentation has the potential to stimulate both an empathetic sharing of feelings and a sympathetic response of being-with someone else's feelings. The fear experienced by a cat when an unfamiliar small child rushes to it with loud exclamations of delight is appresent to us. At the same time, the child's joy and pleasure in seeing a warm, furry thing that is like an animated version of its bedtime toys are also appresent to us. While this mismatch of appresented experiences and emotions may bring a smile to our lips that both recognises the child's pleasure and the familiar response of many (though not all!) cats, we will probably intervene to temper the child's enthusiasm and protect it from a hasty response by the cat. Our *ap*presentation of both the child's and the cat's experience is based on our *ap*perception of each of the other. In this case, the appresentation of the cat's experience or the child's is quite clearly not confused with our own direct experience of the cat or the child. We do not directly share the perceptions or feelings of either, so in this instance we do not really empathise with either the child or the cat though we may sympathise with both.

Alfred Schutz (1971) develops the concept of apperception drawing on both James's and Husserl's use of the term, but it is best understood in terms of his ideas about 'relevance', 'reach' and 'typification' set out in his major work *Structures of the Life-World* (1974). Perception is not random or irrelevant but is structured according to our ongoing motives, interests and actions as we interact with the life-world. 'Thematic relevance' is the flow of attention, usually below the level of conscious awareness, which connects one action to another, directing the attention of visual perceptions to what we are doing. It may sustain a connected sequence of actions, but it can be redirected to focus on something new in the environment, or it can turn inwards to connect musings and thoughts. Thematic relevance may arise from personality and biography and be concerned with motives, wants and projects, or it can be imposed from beyond the person by the actions of others or the contingencies of the environment. If imposed relevance is when information about a financial crisis that will affect the price of food and housing is imparted during a news broadcast, motivational relevance shapes my perception of a programme about food preparation that I seek out and anticipate. The world we are confronted with, both as direct and mediated experience, is shaped by the relevances it has for us.

We can easily ignore or overlook that which is not relevant to us, and we pay little attention to most of what passes in front of our eyes. Some things have an 'interpretational relevance' because they cannot simply be incorporated into continuing action, and we need to stop and make sense of them. The 'whodunnit' of thrillers and crime shows often play with the interpretational relevance of events and objects, both for characters and for viewers. It is striking how important images of a crime scene are for investigators in television programmes, in contrast to textual descriptions or scientific data; they are able to share the interpretational relevance with viewers by simply showing the picture of a character or situation to the camera. Interpretation is not scientific or objective but is motivated by what is relevant to the person (Schutz, 1974, p. 211), and in the unfolding of a crime drama, relevance is oriented to the key characters rather than directly to the viewing audience. Interpretational relevance structures how we look at a crime-scene photo and wonder how the body came to be arranged in that way or with that blood spatter pattern. Thematic relevance blends into interpretational relevance when there is an interruption in the flow of experience that demands reflection on something, a conscious and willed thought that addresses the significance of the meaning of something.

If 'relevance' structures how we see things, so does 'reach'. For Schutz (1974), there are zones of 'actual', 'restorable' and 'attainable' reach in the life-world that refer to the zones in which a human body orients itself to things. The zone of actual reach refers to the spatial and temporal location of a person in the world and what they can reach with their hands and eyes. It constitutes the 'paramount reality' of lived experience that is present for a person who is wide-awake. Restorable reach includes those things that have been experienced before and could be returned to, and attainable reach includes those things that might be brought within the zone of actual reach at some point in the future. If the zone of restorable reach recalls things from biographical experience, the zone of attainable reach is formed by the mediated experiences of other people who tell me, for example, that India is a beautiful country to visit. In relation to the small screen, however, the viewer participates in the zones of restorable and attainable reach available to characters. The viewer is given visual information about where characters have been, and viewers build up a stock of familiar settings for particular characters – occasionally, of course, the viewer may recognise places on the screen that they have been to themselves.

Both relevance and reach structure how we experience things in the world that we know about, and we connect to the unfolding experience of television characters through the same resources. For Schutz, knowledge is stored as 'typifications' that arise from a 'situationally adequate solution to a problematic situation through the new determination of an experience that could not be mastered with the aid of the stock of knowledge already on hand' (Schutz, 1974, p. 231). That is, the stock of knowledge drawn on to apprehend things that we see is organised according to practical interests and experiences, not as a 'code' or abstract system. The availability of typifications means that perception is always apperception; nothing is ever looked at with completely fresh eyes working simply as organs of sight. Put like this, the process of perception sounds personal and idiosyncratic, but our stocks of knowledge are built up through the cultural context in which we live. The typifications that we draw on to apprehend what we see, for example, draw on shared and mediated experiences (for example, images of all types) that have been framed and given value through talk or text.

Schutz extends Husserl's concept of appresentation in his account of the relationships amongst symbols, reality and society when he writes that the:

> ... present element of a previously constituted pair 'wakens' or 'calls forth' the appresented element, it being immaterial whether one or the other is a perception, a recollection, a fantasm or a fictum. All this happens, in principle, in pure passivity without any active interference of the mind. (Schutz, 1971, pp. 296–7)

Even though he is theorising a link between symbols and society, this is not a cognitive process of decoding a formal system of signs. Schutz recognises several orders of meaning that may operate simultaneously and help to explain how the same thing can have different meanings for different people or the same person at different times. Things are made sense of not in isolation but within fields: the physical object in relation to the spatial, temporal and causal relations of nature; the dream object in relation to the order of our dream contents (Schutz, 1971, p. 298).

When we watch television, we engage with the lives of the characters (whether fictional or real) through precisely the same process of appresentation; perceiving them as people with minds who engage with the world in more or less the same way as we do. There is an important difference from the sort of intersubjectivity that Husserl is theorising in that the characters are not 'livingly present'; that is, they are not directly present in the same time and place as our experiencing body. Their presence is, precisely, mediated but they can be more or less 'livingly present' according to the cues as to veracity available through the television screen. A live broadcast (say, of a football match or a concert performance) is more livingly present than a recording, as are people who are appearing as themselves, rather than characters represented by actors. However, the artistry of performance (for example, the skill of actors, directors and camera crew) can also stimulate appresence by being more realistic, less obviously false or unbelievable. The technical qualities of high-definition, high-refresh/response rate video and high-fidelity audio increase the experience of presence, of the aliveness of what is seen through the screen. As the technical form of the mimetic representation is improved, so it enhances the appresence of the experiences that are shown and reduces the intrusion of the process of mediation. There is no reason to suggest that the viewer ever really forgets this difference between their primordial experience and the indirect, mediated experience of the mimetic contents of television. What is more, the discursive narrative devices (commentary, graphics and editing) gloss and interpret what is appresent to the viewer, so emphasising the process of mediation. But those technical features that maximise the veracity of mimesis increase the depth of appresence just as being physically closer to another person, being more familiar with them or sharing a culture with them does. Herein lies the phenomenological crux of television; a mode of intersubjectivity between the viewer and the province of meaning through the screen is achieved – the here-we-are – that demands no complex decoding or linguistic skills because it is appresent to the viewer *as if* they were really there without any reduction to linear, rational, causal or discursive form.

Conclusions

Just as human beings take in what goes on around them as experience that will contribute to the way they make sense of their lives and shape their future actions, so what people see on television has the potential to have this sort of influence. Television creates a sub-universe or province of meaning that is even more accessible and easy to engage with than those that are written or just the product of imagination. This is because the moving images create a continuous present that parallels the temporality of the paramount reality of the viewer. This is often treated as 'realism', but it need not be judged by the standards of reality, provided that the same resources of understanding, recognition, memory and imagination that come into play in paramount reality can be used to make sense of what comes out of the imagery and sound that emanate from the small screen. This process has effects on viewers, but precisely what the effects are will vary according to the specific context of viewing, including, most importantly, what the viewer brings to the perception. The presumption of an unmediated, causal effect is a misunderstanding not only of how television works as a communicative process but also how human beings make sense of the world around them.

The semiological approach to meaning construction may be useful in identifying particular codes and ways of communicating that are incorporated into television programmes, just as they are in paramount reality. An obvious example is the opening title sequence of a programme like *Crimewatch* (see Chapter 1), where a series of signs are used to topicalise the police work of detection and apprehending criminals. However, understanding codes and engaging in a learnt process of decoding is not necessary with television in the same way as it is with the written word. Just because some aspects of television are encoded, and just because there are regularities in the way things are depicted, does not therefore mean that we have to 'read' television as if it were a book or a magazine.

There are two aspects of television that are important breaches of the principle of using everyday life to make sense of reality through the screen. The first is the ideological construction of a form of life that is at some odds with that of most viewers. The lives of most characters in dramas are replete with houses, cars and goods unless the theme is one of poverty and lack of these things. Characters on television, including those in so-called 'reality shows', are more likely to be attractive and appealing than those in everyday life; they have been chosen for this reason. The life experiences that they have are also likely to be more extreme, characterised by death, sexual encounters, thrills and excitement in ways that most viewers' lives are not. There is then a tendency to create a 'televisual' everyday world through the screen that is not a close parallel of the paramount reality of viewers but tends to show either what they might wish for or what they would wish to avoid. There is a second way in which television is not the same as the everyday world around us; it is consciously contrived and designed, not necessarily to have particular moral meanings but rather to attract viewers' attention through the use of aesthetic effects. At one point, Mander encourages viewers to take a 'technical events test' that involves 'counting the number of zooms, fades, cuts, intercuts and uses of music' (Mander, 1978, p. 303). Mander is right, of course; the techniques of 'televisuality', its style as a form discussed in Chapter 4, are important elements of how the small screen constructs other worlds for its viewers.

6
Society and the Small Screen

From telescreen to society

The fear that television might take over society and dominate individuals was articulated in George Orwell's classic *Nineteen Eighty-Four,* first published in 1949:

> The telescreen had changed over to strident military music. It was curious that he seemed not merely to have lost the power of expressing himself, but even to have forgotten what it was that he had originally intended to say. (Orwell, 2004, p. 9)

Modern flat LED, LCD and plasma screens that can be mounted on a wall fit quite closely Orwell's (2004, p. 3) description of the 'telescreen' as 'an oblong metal plaque like a dulled mirror which formed part of the surface of the...wall', but unlike the telescreen, televisions only receive sound and images, they do not gather and transmit them. This means that they do not afford dialogic communication, which, as we will see, restricts their social role; the domestic small screen cannot watch over us as in Orwell's dystopia so, useless for surveillance, the only social effect they have is through the sounds and images they channel into the domestic space. Nonetheless, as Baudrillard puts it, commenting on a 1971 American TV-verité programme on the Loud family, television shows us the minutiae of lives at work and in families through its documentary and reality programmes: 'You no longer watch TV, TV watches you (live)' (1983, p. 53). He argues that television inverts panoptical surveillance by showing us the 'truth' of our lives through exemplary instances – the BBC's *The Family* (1974) and Australian Broadcasting Corporation's *Sylvana Waters* (1982) are other landmark examples.

Michael Apted's *Seven-Up* for Granada Television, which began in 1964 with a series about the lives of fourteen British children who were seven at the time and have been revisited every seven years since, is another model of how the small screen simulates ordinary lives in a reflective mode. Along with soap operas filmed in the studio, the naturalistic family documentary has inspired the 'reality' programming that focuses on families, small groups and individuals' lives. Documentaries have also inspired much television drama that depicts lives, or at least aspects of them, that are similar to those of the viewers who watch them. Rather than our telescreens interfering with our powers of expression, they provide a mimetic means through which viewers can experience and express their own lives.

In this chapter, I want to explore the idea that television both reflects society and extends it. It tries to show us who we are and who we might be, how we are changing and how we should be. Television achieves a moral impact in that what is shown becomes available for people to take the ideas and values of their mediated experience into their own lives – or to reject as unacceptable, the behaviour or norms they have been shown. Although there is also an ideological impact in that these ideas and values may serve the interests of particular groups in society (for example, the commercial interests behind advertisements), where there are different channels and different production companies with different interests, there are opportunities for a variety of voices with different ideas. The overarching interest of all the different voices of television is, first and foremost, to attract an audience, which is how the medium of television conjoins us in the same way that a shared language, a common school curriculum, a legal system and a political structure bring us into being as a society.

I will work towards the idea that the small screen is a very powerful agent of socialisation that keeps the members of society up to date with changes in its moral order. As modernity continues to develop, the rapidity of change increases through technology, economic and environmental crises, political transformations and the violence of war and terrorism, with ever greater demands on the members of society to change their moral outlook. The traditional means of face-to-face socialisation through the education system, the church and the organs of the state are unable to cope with assimilating and disseminating the rapidity of moral change. The media, especially the televisual media that are not limited to textual discourse, are able to respond more rapidly to this process of cultural change. The small screen achieves a form of sociation[1] that is distinctive of later modernity and that brings people into a shared sense of society based on common interests that extends across nation-state boundaries – John Corner calls this 'para-sociality' (1999, p. 95). Clearly, not all interests are shared equally, but values and ideas about how people should and should not act are matters of common concern and constitutive features of society. There are a number of modes through which television communicates these areas of common concern. Firstly, I will discuss how we interact with television, specifically how it addresses us and how we engage with it, and then move on to look at how television fits in with the 'everyday life' of the late modern world. There has been an interesting emphasis on the 'dailiness' of broadcast television, but I want to take a wider view of the pace and pattern of our engagement with the small screen by drawing on Lefebvre's idea of 'rhythmanalysis'. Thirdly, I will argue that television's mimetic capacity for communicating emotional and personal relationships provides its viewers a bridge between their experience of the private sphere and the more abstract public sphere, characterised by political and power relationships; the viewer can take up the role of citizen in relation to the small screen. Finally, I will return to the rather old fashioned idea of 'socialisation' to argue that these modes of engagement contribute to the socialising of viewers into the moral culture of late modern society.

Interaction

Some people, especially those who live alone, like to have the television on 'for company' – the human babble, the laughter, songs, the images of familiar faces, friendly, smiling and rather good-looking can substitute for the absence of other people being in the room. Žižek remarks that the canned laughter on some television programmes may appear to have the effect of reminding us when we should laugh, but he thinks it also laughs for us, acting like the chorus in Greek drama who represent, or stand in for, the audience in the action, feeling the compassion and the sorrow on behalf of the spectators: '... the other – embodied in the TV-set – is relieving us even of our duty to laugh, i.e., is laughing instead of us. So, even if, tired from the hard day's stupid work, we did nothing all evening but gaze drowsily into the TV-screen, we can say afterwards that objectively, through the medium of the other, we had a really good time' (Žižek, 2008). Television is sometimes left on in institutional settings, such as hospitals and care homes, places where people are together but have not chosen each other's company. Some may watch, some of the time, but others may gaze unseeing, using the distraction of the screen to allow his or her own thoughts and memories to be active, to avoid the social obligations of proximate co-presence with others. But watching a small screen can lead to engaging with those through the screen in a quasi-interactive way; laughing, smiling, making remarks out loud and occasionally even shouting or gesticulating at those through the screen. (It is usually politicians that produce this sort of response from me.) How does this interaction with the small screen work?

Joshua Meyrowitz (1985, p. 119) explores the relationship between spectators and those on the screen as 'para-social interaction' (following Horton and Whol, 1956), a form of relationship that allows viewers to regard on-screen personae as their friends. But John Thompson (1995) has thought through in more detail the process of what he calls 'mediated quasi-interaction' to highlight how the situated activity of readers, listeners and viewers changes the nature of their communication with the media. Unlike face-to-face interaction, mediated communication is stretched across either space or time (or both, in the case of newspapers or movies) and narrows the range of symbolic cues available (Thompson, 1995, p. 84). Television is a one-to-many channel of communication across space, which may be 'live' and simultaneous or recorded and separate in time, but (unlike the telephone or radio) gives the receiver visual cues. Nonetheless, the lack of reflexive monitoring of other's responses means that viewers' involvement through the small screen is best described as 'televisual quasi-interaction' (Thompson, 1995, p. 98).

Even though television is a monological form of communication, sending but not receiving messages, the medium can create the illusion of a flow of interaction. Those who make television attempt to reduce the feeling of distance and reduce the impersonality of the communication by using techniques of 'staging' to control the 'speech/image' relations and to make it feel to the viewer as much like ordinary interaction as possible (Corner, 1999, p. 41). The person on screen – the 'talking head' of an expert, news presenter, magazine show anchor or chat show host – will talk directly to the camera, their eyes returning the viewers' gaze, just as if they were in face-to-face interaction, creating what Thompson calls 'direct recipient address' (1995, p. 101). The distance across the room from the viewer to the television screen, the proportionate size of the face and the volume of the voice in the ear all imitate the experience of being at the very front of a co-present audience. The news presenter, for example, is usually in medium close-up with head and shoulders visible, and, as they read from a teleprompter (or autocue), what they say sounds as if it is addressed to each viewer personally; their intonation and facial expression are those we would expect if someone we knew were imparting news to us in the room. Their eyes look straight into the viewer's, and their facial expression, tone and posture are part of the communication process. The chat show host, sometimes reading from a teleprompter, sometimes using cards held in their hand, will also ad lib, creating the particular form of words as they speak, just as if they were shaping them for me, the viewer. Television personalities such as Graham Norton manage to give the impression that they are sharing their feelings of amusement or disgust with us, at the same time as with a studio audience. Unlike the performer on a theatre stage, who must project their voice and exaggerate their gestures to communicate to the back stalls, the person on television is close enough for ordinary communicative expression. Single-channel communication, such as the novel or the radio, does not show the eye contact and simultaneous non-verbal communication available to the viewer of a screen. The co-presence of face-to-face interaction exposes both speaker and receiver to a visual attention that the telephone does not (which may explain many people's reluctance to use videophone communication). But those who appear on television have agreed to, and are usually prepared for, the exposure to both visual and aural scrutiny.

As well as the one-to-one interaction between viewer and presenter, there are other modes of interaction in which the viewer is included. The 'point of view' (POV) shot from a camera of a scene gives the viewer the sensory impression of being present, of seeing and hearing 'as if I were there'. The camera and microphone can work as an extension of the viewer's body in the setting, and the information relayed to the viewer gives a feeling of being there. As Scannell nicely puts it, 'the eye/I of the camera produces not only the effect of being there, but of seeing the scene humanly' (1996, p. 111). The POV is common in documentary and news programmes, but is also used in drama, sometimes intercut with shots of a character whose point of view we are sharing so that it is not only as if we were there, but as if we were actually *in* the body of the character. In the *Spooks* episode discussed in Chapter 4, we shared Adam's point of view at a number of critical moments, and it was made clear by the intercut shots of him trying to make sense of what he was looking at in the dark. This is a perspective that is common in video games in which the visual orientation of the player as a fighter pitted against other combatants is important to the game-play.

Sometimes the POV of a character is presented from over their shoulder so that the viewer is not completely merged with the character but nonetheless for a moment sees things from where they stand – often what is important is their interaction with another character whose responses are seen and heard from a particular vantage point. In actuality programmes we sometimes need a human agent – a reporter or a character – through whose eyes, or at least over whose shoulder, we look so that what we see is modified by what they say. This can produce a compound 'witness' effect as both the reporter and the camera act as witnesses to make the viewer into a witness too (see Chapter 7, pp. 171–6). Interaction in drama and actuality interviews can use 'looking over the shoulder' POV to move from character to character, giving the viewer an intimate sense of different points of view within an inter-action without becoming committed to any one. In reality television, such as Channel 4's *Come Dine with Me* or BBC One's *The Apprentice*, we see the interviewee from over the shoulder of the interviewer, responding to questions that we never hear. The effect is to make us, the viewer, party to a frank and personal discussion about what went on but because we never see the interviewer or hear their questions, we are almost, but not quite, the direct recipient of the answers the interviewee gives. We have seen the scene they are commenting on, but now in a more intimate interaction – in a separate room or the back of a car – we get the participants' reflection on it. Someone's comments directed to a silent and invisible interviewer makes for a style originally developed by documentary makers who wanted to avoid becoming too dominant in the film they were making. The technique makes the viewer almost part of the interaction … but not quite. Other reality programmes, like *Big Brother* (Channel 4; Channel 5), make a feature of interviews in the 'diary room' so the viewer is in the position of Big Brother looking through the Orwellian telescreen while the participants speak directly to the camera responding to the disembodied, invisible voice of the producers.

One common way of showing interaction is for two cameras to take up positions, each at forty-five degrees to two characters as they face each other and exchange lines. In filmed sequences, this can be done with one camera shooting different parts of the dialogue separately. The screen shows each participant in turn in head and shoulders, three-quarter profile, one to the left, the other to the right so that the viewer's position is as a silent third party, turning to hear each contributor to the dialogue. This format, which Thompson (1995, p. 101) calls 'indirect recipient address', is commonly used to capture a dialogue such as that between a visible interviewer and their interviewee or between charac-ters in a drama. In variations on this format, the POV can be adjusted to emphasise one character more than the other or can catch the reaction on the hearer's face rather than seeing each speaker's face. The three-quarter profile is the norm when the speaker is directly addressing someone other than the viewing audience. In a studio setting such as *The Graham Norton Show* (BBC One), the presenter will modulate between addressing the camera and addressing his co-present guests and the audience. The arrangement of the set and sightlines means that as he addresses guests or the audience, he is in three-quarter profile to the camera; we, the viewers, are alternately the addressee and the bystander. This interactive device includes the viewer but reduces the artificiality of their inability to join in; the presenter turns to the guest or audience to relieve the discomfort and formality of direct address. The same technique has been introduced in news and current affairs programmes, such as BBC Two's *Newsnight*, where the interaction between a studio anchor and a reporter provides the viewer with a warmer, less formal interaction than the continuous speech and nearly continuous eye contact of straight-to-camera reading from an autocue.

The ways in which we interact with the small screen produce different modes of sociality that include us as viewers in different ways in the social life on the screen. When we are thrown into the point of view of the camera, people and events are brought close, and the effect can be to make us feel a part of what is going on. The direct address of a presenter or newsreader is rather formal and serious (declarative, informing, making a statement rather than expressing intimacy or exchange), which makes it easier to dismiss or ignore. To create a more intimate sociality, the makers of actuality programmes have developed techniques such as having presenters walk and talk to the camera as if walking and talking side-by-side with a friend. Gestures, references to the scene, turning and talking to someone else and turning back to the camera, even turning to a screen in the studio for a report, help the illusion that we at home are included as co-present participants. This is a difficult effect to achieve and sustain, but when it works, we feel the emotional force of expression that creates an obligation in us to engage and take what we hear seriously; it is meant for us, personally.

Other televisual set-ups give us different degrees of agency so the studio comedy or chat show, the sports event and many other situations put the viewer in the position of a member of the audience – albeit with a better view than most of those actually present. The capacity to move from one part of the arena to another can be vertiginous, such as from the perspective of a line judge close to the action to a viewpoint high at the back of the stadium. But each is a privileged spectator's view, the one close and involved in the action, the other more distant and inactive; the view of a bystander, seeing but unseen. YouTube clips of events like earthquakes and terrorist attacks are often captured as someone was filming a wedding or the scenery of their holiday – the agency, the sense of *being there*, at an awful event, is accidental and unintended but all the more powerful for that.

The degree of sociality prompted by the agency of the camera will alter the moral impact of what is shown. If one sees an act, such as a killing, through the eyes of the perpetrator, the viewer is involved in justifying or explaining it. In the *Spooks* episode discussed in Chapter 4, we see a scene in which a group of Mossad agents kill a man from the victim's view, the killers' view and the 'objective' view of those watching at MI5 via the 'thermal imaging' satellite as well as the perspective of an imaginary bystander watching the scene unfold. The faces of the Mossad agents are masked, but that of the victim who realises he is about to be shot shows fear while 'our' heroes, the MI5 agents, show resigned concern; although we are not in their shoes 'being them', we are 'with them' reacting to their feelings. Thompson rightly argues that overconcern with the text of media output overlooks the 'mundane character of receptive activity' (1995, p. 38). A television programme could be reduced to its script or simply a description of what it shows, but how it addresses its audience is part of the process of communication and will shape the meaning it has. He points out that the different time and space coordinates of a television programme – the move between live studio and distant correspondent and recorded film – are 'interpolated' by the viewer into the 'spatial-temporal frameworks of their everyday lives' (Thompson, 1995, p. 93). To engage with television in this way, and not become disorientated by the rapid changes in place and time, is an everyday accomplishment of viewers and one which shifts their horizon of experience to treat as 'normal' interactions with the past and over large distances as well as with 'real' and 'imaginary' situations and characters.

Everyday life

Audience research has become very popular in media studies, and an ethnographic approach has developed in recent years as a corrective to the simple counting of those who watch particular programmes. There can be value in asking people about what they watch, but the most interesting work has focused on either a particular programme – David Morley's (1980) study of *Nationwide* is a classic that captures much of how television fits differently into different lives – or a particular audience for whom a programme becomes an enthusiasm – here Ien Ang's (1985) innovative study on *Dallas* based on a small number of detailed written responses provides a model. Marie Gillespie has argued that the days of the 'mass audience' are over and suggests that an 'extended' audience engages with connected themes, such as celebrity or sport, across different media. Many of these commentators link being an audience member to the experience of everyday life (Bird, 2003), but it is very difficult to generalise from audience studies to explore how societies use the small screen; there is just too much television and too many different viewers who, while aware of what they watch and why they like it, are unlikely to know how it affects the way they act or relate to other people. Sociologists, Nicholas Abercrombie and Brian Longhurst (1998) have put forward a theory of the 'diffused audience' in which people spend so much of their lives as an audience for various media that they say: '...the media are actually *constitutive* of everyday life' (1998, p. 69). They argue (following Silverstone, 1994) that the ontological security, commonsense understandings, practical knowledge, rituals and symbols of everyday life are integrated with television, especially through the dailiness and habitual watching of it (see also Scannell, 1996).

Even before it can be switched on (see Silverstone, 1994, p. 78), the television as an object came to occupy a role in the everyday life of many households in modernising societies, not only as a transmitter of images but also as an indicator of status in its own right, an 'index of conformity and prestige...a certificate of citizenship...of recognition, of integration, of social legitimacy' (Baudrillard, 1981, p. 54). Despite the proliferation of television platforms, the location of a set in the household living room continues to be a significant piece of furniture around which some of domestic life is oriented for many people. Since the 1950s, the television has competed with the fire as a focus of attention and with the dining table as the locus of family togetherness; in 1956, 15.6 million households in the UK had a set; by 2011, the figure had risen steadily to 27.1 million.[2] As other small screens – television sets in bedrooms, computers and games machines – have drawn some eyes from the family television, this has not yet led to a downturn in ownership or watching of television. In the UK, between 93 and 95 per cent of the population watch broadcast television at least once a week, and something like 75 per cent watch at least once a day.[3] As Baudrillard put it, originally in 1972: 'Lacking a rational economy of the object one deliberately submits to an irrational and formal economic norm: the absolute amount of use in hours. The apparent passivity of long hours viewing thus in fact hides a laborious patience' (1981, p. 55). Before it is switched on, the television set takes up a moral role in the life of modern people as a device that they acquire, situate and attend to along with all the other material objects of late modern lives. Once there, the device becomes insinuated into the everyday lives of most people and becomes a conduit through which their lives are extended through engagement in situations and events beyond the home.

The Marxist philosopher Henri Lefebvre spent much of his analytical energy emphasising the importance of 'everyday life' as the sphere of action in which human beings realise themselves and which was constantly under threat from the forces of modern capitalism. Television is present in the daily lives of most of the population of modern societies. It is an object of everyday life, in Henri Lefebvre's terms, an item of household equipment that, like the cooker, the refrigerator and the washing machine, is used routinely and habitually, fitted into the patterns of ordinary living, largely unquestioned or remarked upon. The 'small technical actions' of switching on, choosing a channel and getting the volume right 'intervene in the old rhythms rather like fragmented labour in productive activity in general' (Lefebvre, 2002, p. 75). But 'leisure machines' such as the television have an 'extra-technical' meaning and character as they chop everyday life up, leaving margins and empty spaces, increasing passivity (Lefebvre, 1991, p. 32). We wait for the programme we want to watch, we rush the completion of a meal to fit with the schedule, postpone other activities and risk boredom as we give ourselves up to watching. The inclusive 'we' implies that everyone integrates television viewing into their lives in the same way which is clearly not true, but watching television in the family home is a culturally established pattern in northern societies; in other cultures, television may more often be viewed collectively, in public spaces, even in the open air. In the living spaces of the home, television has brought with it patterns of behaviour that are as routine as they are mundane, a passive, non-participatory 're-privatization' of everyday life. The viewer is overwhelmed with news and information in a 'great pleonasm' as the realm of the everyday is presented and re-presented endlessly; 'the mass media have unified and broadcast the everyday' (Lefebvre, 2002, p. 75). Programme makers strive to attract viewers with the exotic and the extraordinary – but they also need to connect with viewers' everyday experiences. Watching television is a part of everyday life, but the unfolding of everyday life is also the main content in soap operas and reality shows, news and documentaries, dramas and domestic comedies.

The everyday lives of ordinary people are too familiar and banal to make entertaining viewing, so one popular device for making what Baudrillard calls 'TV-verité' (1983, p. 49) is to turn it into a competition. The presentation of people who have not been trained as actors or presenters and who have no formal script has come to be known as 'reality television', so called because of its derivation from the distinctive practices of television documentary making (see Corner, 1996). This mode of television is always contrived to exaggerate and amplify the ordinary, offering its viewers the excess and extremes of everyday life. For example, in the Channel 4 series *Come Dine with Me*, five ordinary people (neither cooks nor experienced performers) compete for a cash prize by cooking a meal for each other on different nights in their own homes. Each judges and scores the host, away from the other contestants, 'privately' in the intimate space of the back of a taxi or a quiet room in the house – but with the million or more viewers watching them. The programme's viewers can enjoy how the elements of the various cooking shows (new recipes and menus, skills in preparation, techniques and technologies of cooking) are brought together with elements of 'reality' television that put people together competitively to see how their personalities cope with the situation and each other. The everyday life activities of eating and talking in people's homes are reproduced, but the show nicely captures the way dinner parties are a break from the quotidian flow, a special occasion with dressing up, alcohol and an extra effort made over food preparation that requires different performances from host and guests. Because viewers can only imagine what the food tastes like, the real topic of the programme is how the contestants feel about each other and the food – and how they show their feelings in interaction. The predominant mode of indirect recipient address means that the television viewer is an onlooker, and it is easy to be critical of the performances – of how the food looks, how people behave, even how they dress and their homes are prepared – and imagine how they might do better.

Television programmes about food might be linked to the promotion of excessive eating or alternatively to trying to counteract obesity or poor diet (Inthorn and Boyce, 2010), but it is difficult to understand a programme such as *Come Dine with Me* as a blunt ideological instrument from which viewers learn by watching and imitating the values and activities that they see. The programme is built on the cultural values of holding a dinner party (of giving pleasure to guests through food, drink and sociability), and its competitive element is ostensibly about finding the best way to realise these values. But much of the viewer's pleasure is in seeing how contestants' different performances of hosting and cooking work out. As the series progresses, it becomes clear that as a competition there are more different ways of being 'good' at hosting a dinner party than there are of winning the lottery or running a race. As with most 'reality' television, the topic is the mores of everyday life – how to get along with strangers, how to prepare food, how to eat in company, how to be a guest in other people's homes, how to express pleasure or dislike appropriately and, indeed, how to cope with the effects of alcohol in social situations. There is always a risk in shows such as *Come Dine with Me* that one class or gender is displayed for amusement and disparagement (Skeggs, 2005; Skeggs, Thumin and Wood, 2008; Tyler, 2008). But the participants are from many different classes and regions, professions and interests, and the programme is not about abstractions or principles; there are no demonstrations of excellence or expert judgement.

Television watching is itself a mundane activity that meets ordinary needs for amusement and distraction as well as an interest in how the rest of society gets on with ordinary life. It is through ordinary needs that people are brought together in the mundane interactions that make up the stuff of everyday life, and for Lefebvre, even in the face of alienation, '...these needs in everyday life are a cohesive force for social life even in bourgeois society, and they, *not political life*, are the real bond' (1991, p. 91). It is through resistance to the institutional forces of capitalism that the sphere of everyday life can be the site of a Marxist-inspired critique of contemporary society in which human beings realise themselves and transcend the separation of the individual from his or her self. As viewers watch television, insofar as they recognise the connection between the situation of the characters – be they played by actors or 'real' people – with the conditions of their own lives, then they are participating in a critique that questions both the lives played out on the screen and their own lives. This does not need to be a conscious process of thinking, of rationally weighing up criticisms but when watching a programme like Come Dine with Me it may be expressed by the viewer along the lines of 'That looks awful food', 'That was very tactless', 'How rude'.

Another way in which television fits into everyday life is to both reflect and provide a structure for its temporal rhythms. As Silverstone (1994, pp. 18–23) recognises, the broadcast of news follows a regular pattern that can structure the day and provide the viewer with a habitual and familiar routine. Paddy Scannell (1996, pp. 144–78) suggests that the ontological nature of the broadcast media as 'sociable' involves the temporality of 'dailiness'. Broadcast news emphasises clock time, picking out midday, six and nine o'clock, connecting the listener or viewer with the culturally shared spacing of time, coordinating their lives with others. Williams's broadcast 'flow' (see Chapter 4, pp. 87–9) includes many programmes that are thematised to fit with the zones of the day they are accompanying – breakfast time, morning, midday, afternoon, early evening, mid-evening and late-evening. There are different viewers (children, adults) and different activities (preparations, meals, relaxing) that fit with these different time periods. Scannell's idea of dailiness includes not only the linear sequence of zones but also the cyclical reoccurrence of this sequence everyday and the cyclical recurrence of seasons – such as Christmas, the football season, the summer recess of Parliament – all of which are reflected in programming. Television series usually have a weekly cycle of episodes and often also have their own annual 'seasons'. The many forms of televisual seriality (see Chapter 4, pp. 90–2) mean that there is a regular return of characters and settings, of themes and issues, of televisual styles and incidental music, that give a rhythmic quality to the small screen's role in everyday life.

The cyclical patterns of everyday life are part of Henri Lefebvre's (2004) method of rhythmanalysis, which is itself a counter to the reification entailed in a static analysis of 'things' (products, commodities, institutions, matter). But for Lefebvre, the media and the mediatisation of everyday life are rather more problematic than they are for Silverstone and Scannell. He argues that the flow of the media is continuous throughout the world, running through the night: 'At any given hour, your instrument can fish for a catch, a prey, in this uninterrupted flow of words in the unfurling of messages' (Lefebvre, 2004, p. 46). Twenty-four-hour news channels ensure that this is the case for television now in a way it was only true for radio for much of the 20th century. But what is there to catch? What meaning does it have? Lefebvre distinguishes between what the media 'present' – the filling of *this* moment in time – and 'presence', which the media can only simulate: 'With presence there is dialogue, the use of time, speech and action. With the present, which is there, there is only exchange and the acceptance of exchange, of the displacement (of the *self* and the *other*) by a **product**, by a simulacrum' (2004, p. 47 – emphasis as in original). The liveness of televisual quasi-interaction can mark the 'present', but it cannot achieve 'presence'. What is made present of the spatial separation that television bridges is a commoditised, manufactured product, not the presence of life itself.

But the viewer is not simply a cultural dope, forced to consume like an empty vessel being filled up with product. The viewer will have an *internal* dialogue with the content of the small screen, agreeing disagreeing, enjoying, being bored by, judging, questioning and so on. That internal dialogue may, of course, be shared with others, such as those also watching, or in everyday conversations about television content with friends, workmates or others. The practice of everyday life can itself become critique when it becomes praxis, that is, when it involves not just the routine habitual flow of actions through which the living being survives but includes reflection on, and questions about, the determinants and necessity of that flow. This praxis may include the watching of television, since it offers endless material for provoking reflection and critique in the watcher. For example, one feature of *Come Dine with Me* is the commentary delivered in a jocular voice by an invisible man who gives background and continuity to the scenes but also gives ironic, often quite acidic, comment on the actions and on-camera remarks of the contestants. He says out-loud words that often crystallise or anticipate the viewer's reaction to the contestants, providing an internal voice of the 'viewer/camera', a sort of silent quasi-interaction with contestants (they cannot hear and do not reply) that mimics the internal silent 'dialogue' that viewers often have, commenting on and criticising what they are watching. The commentator is, of course, part of the production that has edited and shaped the programme, but in juxtaposition to the actions and comments of the contestants he legitimates the viewer being judgemental and even dismissive. When the viewers comment to themselves or to other people, it is not a true dialogue, because there is no exchange back and forth, but it is dialogic in the sense that the commentary is a direct response to the speech and action of characters that, if they could hear, they would respond to in turn.

Public sphere

The concept of the 'public sphere', introduced to the social sciences in 1962 by Jürgen Habermas (1992), gives a key role to the media – early print media – in establishing a civil society where those who do not hold positions of state power can share information and debate what should be done. The public sphere is a social space in which deliberation leads to the expression of shared opinions outside the formal decision-making process by people who may in other respects have conflicting interests. This expression of public opinion may then feed into the discussions and decisions of government, making the public sphere a precious institution in a democracy. Political parties and governments in England, France and Germany claimed the influence of public opinion as articulated through the informed and critical debate of various journals of the 18th century (Habermas, 1992, pp. 57–88). The public sphere that Habermas describes grew out of the literary societies, salons, coffee houses and *Tischgesellschaften* (table societies) in which there was informed discussion in 18th-century Europe. The character of this debate is important because it disregards social rank and establishes issues of common concern as proper matters for cultural exchange. The public sphere was inclusive in principle: 'The issues discussed became "general" not merely in their significance, but also in their accessibility: everyone had to *be able* to participate' (Habermas, 1992, p. 37). Of course, those who actually participated were the male members of the bourgeoisie; the professional and merchant classes who were not only property owners but also literate and educated. The cultural goods of the public sphere – newssheets, journals, moral weeklies, books, theatre performances, political pamphlets – were in principle available to all but in practice only to those who could afford and understand them. The articles and letters published in the journals were topics for face-to-face interactions in the increasing number of coffee houses and other meeting places, but the 'moral weeklies' (that included the *Tatler*, the *Guardian*, and the *Spectator*) were concerned with promoting self-enlightenment and self-understanding. Habermas describes the themes of charity, education, poverty, gambling, fanaticism, pedantry and tastelessness and how Addison, one of the founders of the *Tatler*, 'worked toward the spread of tolerance, the emancipation of civic morality from moral theology and of practical wisdom from the philosophy of the scholars' (1992, p. 43).

For Habermas, the public sphere is an opportunity for critical-rational debate that has the potential to lead to enlightenment – in a philosophical rather than spiritual sense – that he associates with Kant's concept of pure reason. It is a social space of thought and argument shaped by a morality of debate that Habermas also derives from Kant, in which the political expression of a shared will in the public sphere is oriented to freedom; it is precisely not about cunning or domination, nor is it constrained by rules or institutions of justice. The private sphere is where the bourgeois individual can express personal opinions and expect them to be simply accepted by family and friends, but in the public sphere they will be subject to question and criticism, and the sum of different opinions will cancel each other out, leaving only those that are generally agreed and accepted. Now Habermas argues that there was a 'structural transformation' as the relationship between public and private spheres began to change towards the end of the 19th century. On the one hand, the private life of domesticity and leisure became more distinct from work life as people increasingly worked away from the home in collective organisations. Work became regulated and controlled in terms of time and space, and even the bourgeoisie of merchants and professionals spent less time together discussing common interests. On the other hand, public life became more private as those potential contributors to the public sphere retreated into commodity consumption for themselves and their families – consumption here included the media and the electronic and broadcast media in particular: '... rational-critical debate had a tendency to be replaced by consumption, and the web of public communication unravelled into acts of individuated reception, however uniform in mode' (1992, p. 161).

For my argument that television is a key medium through which the moral order is shared and modified, it is important to recognise it as contributing to the public sphere. Television does not replace the print media of the 18th century, but it is less dominated editorially by financial and class interests than contemporary print media. The variety of materials available on television, intermixed in the mélange of the flow created by viewing choices, makes available to viewers a range of ideas and possibilities about how people might and should act. Habermas saw the mass media as undermining the public sphere, whereas I want to argue that they have opened up civil society in new ways – the very process that he sees as restricting the public sphere, I want to argue extends it. He writes: 'The deprivatized province of interiority was hollowed out by the mass media; a pseudo-public sphere of a no longer literary public was patched together to create a sort of superfamilial zone of familiarity' (Habermas, 1992, p. 162). This follows a familiar Frankfurt School critique of the cultural industries that harks back to a purer bourgeois culture, in this case, one that was 'literary'. This dazzling display of distaste that a 'world of letters' should have been displaced by a 'sphere of culture consumption' is reminiscent of Adorno and Horkheimer's (1979) criticism of the mass deception perpetrated by the culture industries; just by changing the channel of communication from that preferred by the traditional cultural elite to electronic channels that are cheap and accessible enough for everyone, somehow the content of communication is apparently debased. The media, even literary media, are unable to continue to express public opinion because they become oriented to the mass and are dismissed as: 'canned goods', 'ready-made, flexibly reproduced, barely internalized, not evoking much commitment', 'small talk', 'publicity staged for the purposes of manipulation or show' (Habermas, 1992, pp. 246–9). News is dressed up as entertainment through the personalisation and sentimentalisation of politics and cynicism towards institutions, which effectively neutralises criticism of public authority. Habermas claims that 'the world fashioned by the mass media is a public sphere in appearance only', while 'the integrity of the private sphere which they promise to their consumers is also an illusion' (1992, p. 171).

There is a literary fetishism at work here that recurs in critics of television such as Jerry Mander (1978) and Neil Postman (1985) in which writing is considered to be a purer form of information than audio-visual media. The experience of reading is different; it is a very private mode of consumption in which the quasi-interaction amongst reader, characters and authors is internal and is less easy to share or to integrate into other activities. (Ironing while reading is much more difficult than while watching television.) For some purposes, this may be the case, and indeed, critical-rational debate is not the characteristic form of the great majority of what appears on the small screen – although this does not mean that the medium cannot offer such debate to its viewers. What I want to argue is that the phenomenological form of the medium of television offers a *different* form of the public sphere, one that is better able to communicate feelings, emotions and practical consequences. Even more importantly, those included are not restricted to certain sectors of society, and anyone can join the viewers who listen in and watch. Television programmes can promote discussion and sharing of views and opinions in a variety of ways – not necessarily deliberative – that are important for stimulating the collective sense needed for cohering society as a moral order. Habermas's arguments about the public sphere – its form, historical emergence, transformation and its decline – have been the subject of vigorous academic discussion (see, for example, the collections by Calhoun, 1996; Crossley and Roberts, 2004; Wodak and Koller, 2008). The mass media have also been argued to be contributing to citizenship and the contemporary and 'suburbanized' public sphere (Dahlgren, 1995; Stevenson, 2003; Silverstone, 1994). Habermas's own later work retreats into the philosophically fascinating questions about the ethics of discourse and how the generation of social norms can be judged to be free of distortion (1990). Sadly, Habermas doesn't clarify his view of the role of the media, but it is difficult to see how at least serious news and discussion programmes on television could be excluded from the public sphere. Rather than enter this complex debate, I want to propose that television not only contributes to Habermas's version of the public sphere but also can extend it in ways that Habermas does not discuss.

The idea of *deliberation* is central to Habermas's concept of the public sphere and involves reasons given to support opinions, criticism of those opinions and argument that challenges reasons. This is very different from the invocation to be and act in particular ways that had been characteristic of the pulpit and the hustings, where the authority of God or personal standing were the only reinforcement necessary. Participants in deliberation each have to be persuaded and have, at least in principle, the right to question what they hear or read and present their own views. There may be many sides to an argument and many factors to be weighed, and interpersonal interaction is the ideal format for deliberation. However, the quasi-interaction of television can begin to approximate and involve millions rather than dozens. *Question Time* (BBC One), a regular television series begun in 1979 that is a television version of *Any Questions?*, a radio show running since 1948, has the format of a panel of four people, three of whom are usually politicians, answering prepared questions from an audience. The panellists on the show (which has led to other versions in other countries including Australia) respond to the questions on topical issues with opinions and reasons and comment on and criticise the comments of other panellists. A chairperson controls who can speak when and for how long and will try to ensure that those criticised have a right to reply and periodically invites the audience members to contribute views. Occasionally, at the end of a discussion, the audience will be invited, on a show of hands, to support or disagree with a particular view to gauge opinion in the room.

This format of a chairperson and a panel of a small number of people who have differing views is a regular feature within programmes such as *Newsnight*, *Channel 4 News*, review shows and some documentary or investigative programmes. The chairperson 'intermediates' introducing speakers, putting topical questions, moving the discussion on, managing the turns between speakers and eliciting opinions and views. The chairperson takes the role of the viewer's agent, not simply seeing fair play but also ensuring that questions are answered and that what is said is clearly expressed. It is the requirement of clear expression that limits who is likely to be a participant in such public deliberation; those asked to take part are usually experienced and sometimes trained in the art of speaking in this semi-formal situation with an audience. The skills needed are partly rational-critical, but they are also rhetorical because the presentation of the argument, the facial expression and demeanour, the tone of voice and the choice of words all affect the power of what is said. Politicians, journalists and public relations people are particularly well prepared for this role, but experts, particularly academics, and the leaders of pressure or special-interest groups will have gained the skill of deliberation in a variety of face-to-face contexts, including seminar rooms and public meetings. These participants in public debate on television are likely to be bourgeois or middle class in that they have been educated or trained, although some, such as trades union leaders or special-interest group representatives, display a very effective mixture of talent for expression and passion for their views.[4]

In Habermas's public sphere, those who participated were a small fraction of society; upper middle class men in the cities. Even amongst this elite group, it is likely that some people spoke more often or whose views were more highly respected and so listened to. Some will have been better educated, some will have been better informed, and some will simply have been better at arguing their case. Habermas offers no detail on who actually participated in deliberations in the coffee houses and no account of the rhetorical or discursive strategies they used – how would he know? Personalities, as well as reasons and opinions, will have shaped debate, and some will have retreated from speaking while others will have come to the fore as influential opinion leaders (see Chapter 5, pp. 97–9). In any public sphere, whether in a coffee house or on television, the tongue-tied are not heard, though their cheers of support may empower another speaker to effectively speak for them. Business and professional interests will enter into discussion and understanding – and will indeed often set the topic for debate. Even where there is no stage, podium or pulpit, face-to-face interactions involve persuasions that are not reducible to the rational-critical or even the rhetorical, which is why Baudrillard (1994) rightly talks of the 'staging of communication' to refer to the way meaning is managed in the mass media. Of course, if a debate is so stage-managed that the participants are no longer deliberating – rather as happens at contemporary political conferences – then the public will lose interest. Just like those who wrote the newssheets and informative documents for the public sphere of the 18th century, broadcasters are interested in creating debate that is lively and informed and represents a range of opinions because they want to attract an audience.

Although television can communicate deliberation, what is distinctive about audio-visual media is their capacity to communicate emotion and feeling. Kay Richardson (2008) explores the difference between the traditional radio format of *Any Questions?* and a contemporary television talk show *Sally* in which the topic at issue is more likely to be personal behaviour, and interactions do escalate into confrontation as emotions, rather than reasons, are expressed. As she says, 'The entertainment orientation of the daytime talk shows does not necessarily make their contribution to the public sphere negative or irrelevant' (Richardson, 2008, p. 396). The rather different mode of the public sphere characteristic of a series of talk shows including *The Oprah Winfrey Show* (King World/CBS – 1986–2011), *The Jerry Springer Show* (NBC – 1991 – present) and *Kilroy* (BBC One – 1986–2004) are not deliberative and not usually about rational-critical discussion of political issues, but they are precisely about matters of morality; sexual relationships, family relationships, obligations, behaviours, attitudes and actions.

In a show like *The Jeremy Kyle Show* (ITV1), the presenter manages interactions between guests in a way that promotes verbal aggression and the expression of emotions and is not shy about offering moral judgements on their behaviour. The telling of deeply personal stories and the outpouring of feelings happens on a stage in a studio with an audience who, from time to time, have the opportunity to offer their views if only through a collective utterance of shock, amusement, approval or disapproval. The camera angles often include the audience as a background to Kyle as he moves around the studio in a style that Phil Donahue began in the US in the 1970s, and Jerry Springer and Oprah Winfrey developed in the 1980s and 1990s. These highly staged shows are at the extreme of 'car crash television', in which the viewer looks on with fascinated horror at the lives of others where moral failings that have brought anger, pain and unhappiness are exposed for all to see. The situations that the panellists discuss and confront each other with may be personal but they are exemplary of what goes wrong when people lie, cheat, are disloyal and hateful to each other. Jeremy Kyle's skill as a presenter is not only to facilitate their public confessions, by using what the programme's researchers have discovered, but also to summarise their moral limitations and tell his guests what they should do. His guests tend not to be articulate, educated, successful or well off and it is easy for him to catch out the contradictions in their stories – they are not experienced at putting their case cogently or using reason to support their views. The guests swear, gesture and shout over each other and do not recognise the moral principles of argument or reasoned deliberation, but they do, surprisingly, accept Kyle's irritable teacher moralising. He is against violence and emphasises the values of loyalty and the priority of children in a relationship.

Habermas is criticised for the elitism of his preferred version of the public sphere that excludes the non-literate and uneducated classes, it excludes the intimate sphere of the household including gender relations and seeks consensus in a way that excludes extreme, pluralist and irreconcilable positions (Fraser, 1996; Garnham, 1996). Nick Couldry offers an interesting approach to the idea of the public sphere as providing 'feedback loops' that help to generate 'categories' in a Durkheimian sense that enable viewers to make sense of social reality (2008, pp. 78–9). The feedback loops are how viewers make sense of the world about them by reference to what they learn from the media and how the media also makes sense of the world by reference back to what the media have shown. So, guests on *The Jeremy Kyle Show* act as they have seen characters in a soap opera act, and characters in a soap opera act like those on *The Jeremy Kyle Show,* and both references are used by viewers to make sense of reality. Couldry is not suggesting a crude imitation of behaviour (though this may, of course, happen) but that the social 'categories' displayed in the media – on *The Jeremy Kyle Show* these might include the nagging wife, the unfaithful husband or bad mother– are 'associated with the gradual emergence of new categories which help to organise everyday action and thought' (1996, p. 79). This implies that any distinction between the media and everyday life is constantly threatened as the categories keep confounding the boundary, as summed up in the contradictory notion of 'reality television' (Couldry, 2008, p. 83).

Habermas was concerned with the morality of political deliberation, but the public sphere is even more important as a realm in which morality itself is shaped. Critical rational deliberation happens in documentaries, news, magazine programmes and debates and panels, but it includes the non-discursive or quasi-discursive expression of human values of gesture and moving image that literary communication cannot carry. But television *also* provides a public sphere for the discussion of private lives and the morality of interpersonal and emotional relationships. Democratic society is not only about the decisions made about money and war, it is also about the freedom to express identity and engage in intimate and private lives in a way that is accepted and approved of by the culture as a whole. Television is a very different medium from the journals that were read in the 18th century, but as small screens have become portable, programmes can be watched in public spaces, including buses and trains, waiting areas and even in pubs and coffee houses. Television brings the public sphere into the private, and the private sphere into the public.

Socialisation

The idea of socialisation is normally associated with the induction of children into the culture of their societies, through the family (primary socialisation) and through education (secondary socialisation). What children should be allowed or encouraged to watch is a contentious issue, which suggests that television plays a part in their socialisation. Television can be an educational medium that supplements other channels of socialisation, although its quasi-interaction is an inadequate substitute for the full communicative interaction with peers, parents and other adults, especially teachers. Rather than enter the debate about children and television, what I want to argue here is that television is a medium that socialises *adults* into the rapidly changing moral values and mores of their society. Socialisation is never complete, especially when society is continually changing; the culture of late modern societies, enhanced by electronic media, is in a constant state of flux. Its boundaries continually expand through travel and communication, so that ideas, practices and ways of being are exchanged over ever greater distances. The term 'socialisation' has fallen out of favour with sociologists in recent years (social psychologists and educationalists continue to use it), but in the middle of the 20th century it was a key idea, especially in the dominant American functionalist tradition. Ely Chinoy, in a 1960s standard textbook on sociology, writes that socialisation serves two major functions:

> On the one hand, it prepares the individual for the roles he is to play, providing him with the necessary repertoire of habits, beliefs, and values, the appropriate patterns of emotional response and the modes of perception, the requisite skills and knowledge. On the other hand, by communicating the contents of culture from one generation to the other, it provides for its persistence and continuity. (Chinoy, 1961, p. 75)

While he doesn't use the term 'moral order', you can see that Chinoy is describing much the same process as that which I've called 'morality'; the acquisition of values and habits appropriate for the society, the passing on of appropriate ways of acting and responding. The functionalist idea of socialisation emphasised face-to-face interaction and saw the family as a 'system' that created personalities who would then be able to fit into society. Talcott Parsons, for example, believed that human personalities are made, not born, and he described families as '"factories" which produce human personalities' (1956, p. 16). The mass media were regarded as sources of information and entertainment rather than moulders of personality, and although there was concern that television might interfere with 'normal' socialisation through the family and education, there was little sense that it might be a medium of socialisation in itself.

As the functionalist view of society as a set of interconnected systems fell out of favour, so the concept of socialisation declined in importance, but it also played a part in a very different sociological theory that has persisted. The pragmatic and phenomenologically inspired sociology of Berger and Luckmann's *The Social Construction of Reality* (1966) saw socialisation as a dialectical process between the emerging individual subject and the social reality that they are born into. The individual subject's social reality is derived through interaction with significant others (primary socialisation; usually in the family) and through induction into roles (secondary socialisation; especially work and family roles). They saw social reality as emerging from those shared habitual and routine social actions that become institutionalised and sedimented into traditions and roles. The aspects of reality that the subject creates with the others around them are to do with the flow of everyday life; practices like going to work, maintaining religious beliefs and conversations with a spouse. The social subject has their own 'subjective social reality' that becomes progressively more symmetrical with the 'objective social reality' that is beyond his or her bodily sensations and biographical experience. But because the individual cannot internalize all that is objective reality, and subjective biography is never fully social, the dialectical tension between subjective and objective reality remains. The two versions of social reality are never completely symmetrical, never coextensive. Berger and Luckmann describe the relation between these two realms as being continually produced and reproduced 'like an ongoing balancing act' (1966, p. 154).

What is striking, despite being written at a time when mass communications were taking off, is the emphasis in Berger and Luckmann's quite radical argument on direct, co-present, social interaction. The role of the media is hardly mentioned; the impact of newspapers, novels or movies is not regarded as of any significance in the creation of social reality, and there is surprisingly little discussion even of the knowledge acquired in formal education. Douglas Kellner remarks on the absence of interest in the mass media in Berger and Luckmann's account of socialisation, pointing out how television 'plays a central role in socialization from cradle to grave' (1990, p. 127) and so 'is a force for cultural homogenization and blandness' (1990, p. 126). It is difficult to see why television is any more of a force for cultural homogenization and blandness than a dominant religion or state education, but during the 20th century the media became an element in the everyday routines that maintain social reality. Berger and Luckmann do describe a realm of 'symbolic universes' that are constituted by ideas, myths and religions and that provide 'sheltering canopies over the institutional order as well as over the individual biography' (Berger and Luckmann, 1966, p. 120). Symbolic universes order and legitimise both objective and subjective social realities, and it is in this dimension of social life that both morality and mores would be found. Although religions provide the model of symbolic universes for Berger and Luckmann, modern pluralistic societies have multiple 'partial universes' of beliefs and idea systems that coexist uncomfortably with rival advocates for their different versions of social reality (for example, religion, lifestyle, health, science, community engagement, politics and protest).

The small screen is a potent communicative channel for these 'partial' symbolic universes (or 'sub-universes' in James's terms, or 'provinces of meaning' in Schutz's terms – see Chapter 5, pp. 107–9) that can provide viewers with a continuous type of tertiary socialisation into society's evolving objective reality as the subject progresses across their life course. The socialisation of adult viewers through television is not the same as the instruction, teaching or advice through which children learn. As a channel of communication that, at best, offers quasi-interaction, the small screen provides a form of vicarious experience through which the appresentation of situations can be engaged with, without any of the actual consequences that would ensue if the viewer were really experiencing the situation. The unfolding action on the screen means that it is easy for viewers to sympathise and empathise with the situation of characters and even imagine themselves as the adulterous lover or the vengeful murderer. They can see what the outcome might be for a line of action, they can see the reactions of other characters, and they see an unfolding set of consequences, both in fictional and actual representations of life. The social life portrayed on a screen represents a possible objective reality while characters, both fictional and actual, utilise a subjective reality that guides their actions; the viewer can see the extent to which actions are plausible and how characters collectively construct a viable shared version of reality. This does not mean that there is any reason to adopt or copy the behaviour. The televisual experience of exploration at the level of the imaginary may have the opposite effect of clarifying why a line of behaviour is inappropriate. The habits, conduct, mores and morality of a possible world are played out on the screen in a way that puts the viewer in the position of an 'impartial spectator' (See Chapter 8, pp. 199–205). Contemporary television programmes seldom contain simple moral tales, as *The Lone*

Ranger did; heroes are flawed, villains have redeeming characteristics and unintended or unanticipated consequences are revealed as lines of action unfold. Late modern viewers become bored if the moral dimensions of programmes are too black and white, which is why programmes like *The Sopranos* (HBO), *The Wire* (HBO) and *The Killing* (*Forbrydelsen* – Danish Broadcasting Corporation) that explore morally transgressive characters, have had both popular and critical success as post-millennial television series with wide DVD sales.

Although the role of the media in the process of socialisation is seldom recognised, Dennis McQuail's standard textbook says:

> ...the media can teach norms and values by way of symbolic reward and punishment for different kinds of behaviour as represented in the media. An alternative view is that it is a learning process whereby we all learn how to behave in certain situations and the expectations which go with a given role or status in society. Thus the media are continually offering pictures of life and models of behaviour in advance of actual experience. (McQuail, 2005, p. 494)

If the structural-functionalist account of socialisation tended to 'oversocialise' human existence, reducing the social subject to an empty vessel to be filled with the norms established as functional for the smooth running of society, there is a risk of an 'undersocialised' version of human existence in which the potential for social actors to shape their everyday lives according to their own interests is overemphasised. The small screen provides a complex of symbolic universes that interplay with everyday life in a process of continuing socialisation.

Adults are able to choose what they watch. Broadcasters may plan the flow with the aim of persuading viewers to stay tuned, but where there are multiple channels and platforms, viewers will make their own flow for a viewing session, intermixing programmes of different types from different sources (broadcast, DVD, DVR, Internet) even while they watch the same screen. The range of programmes broadcast and circulated for the small screen addresses many categories of people – this breadth of content and potential audience is one of the things that makes audio-visual media so difficult to analyse yet so important sociologically. Within the private sphere, individuals and households have a high degree of choice about what to watch but too great a divergence between the values, ideas and interests of the programme and those of the person will lead to their decision to watch something else or nothing at all. Television often rhythmically repeats similar themes, issues and ideas, especially within a series or a genre but within familiar formats, programme makers strive to stimulate the viewer's curiosity and attract new viewers. The introduction of homosexual relationships into long-running soap operas, for example, has both built on the values of characters established over weeks or months and also disturbed taken-for-granted ideas about other aspects of personality. The programme makers will allow different characters to react in a variety of ways to reflect the range of responses expected amongst viewers; if some reactions are slightly exaggerated, then it is all the easier for viewers to either follow or distinguish their own reaction from those of the characters.

For example, during 2006 *EastEnders* character Honey gave birth to a baby, Petal, who was identified as having Down's syndrome. The storyline developed over a number of months as Honey was distraught and found it very difficult to accept Petal as she was, while her father, Billy, seemed overjoyed to have a daughter and was keen to get involved in bringing her into the world. Other characters reacted in different ways as Petal's condition became known; some enthusing and sympathetic, others trying to ignore the syndrome and its potential consequences. The storyline got a critical and defensive response from representatives of professional organisations, but it had been carefully researched, and the broadcasters clearly intended to use the vehicle of the soap opera to explore a difficult issue that affects a small but significant number of people. (About 750 babies with Down's syndrome are born each year in the UK). As the makers of *EastEnders* and many other programmes recognise, television is a medium that can socialise adult viewers by raising topics and presenting a variety of views, opinions, ideas and values. This was an instance where programme makers used the medium of television deliberately as a means of adult socialisation and shows how the medium does not lend itself to crude propaganda because the viewing public – including those with a particular interest in the content – expect a degree of subtlety in the presentation of moral issues that does not simply reflect a single set of values or moral code.

An approach to moral education that does not rely on particular values or a moral code was what Durkheim (2002) proposed in a set of lectures on education at the Sorbonne in 1902. His message was that moral education should be undertaken in a very different way from how religious teachers and 'moralists' approached it. What was needed was to teach young people how to engage with the moral order, recognising it but finding a way of responding to it, keeping open the possibility of change. He disagreed with those philosophers – he cited Kant, Mill, Bentham and Spencer – who had attempted to identify an underlying rule from which to generate a moral philosophy (see also Durkheim, 1974) and wanted the moral order to be dynamic and flexible. The teacher of morality must, as he put it, 'guard against transmitting the moral gospel of our elders as a sort of closed book' (Durkheim, 2002, p. 13). He saw science, rather than religion, as the symbolic centre of the moral order and wanted to promote rationality and reason rather than a set of 'dos and don'ts' as its core values. Instead of imparting an ethical code, the role for teachers was to enlighten their pupils about the social origins of morality. In modern, organic societies, along with rationalism had come an increasing individualism, and he argued that each individual had to learn to question the moral order from their own perspective.

The viewer of television is confronted with a continual stream of programming that, whatever else it does, shows moral content which demands some response; can the moral reprehensibility of this murderer's actions be ameliorated by his past circumstances; is the kindness of a character in one context outweighed by their selfishness in another; does the footballer's skill and passion mitigate his 'professional foul'? As the viewer becomes caught up in the unfolding events on the small screen, their interest in what happens is connected to the morality displayed and the responses of other characters to it. Television programming – particularly for adults – eschews the didactic mode of moral instruction but offers the opportunity for both rational and emotional reflection on how actions and events lead to certain outcomes. The *EastEnders* Down's syndrome story did not have a simple message beyond raising the issue and showing a number of responses, including the consequences of those responses for other characters. Just as Durkheim cautioned against teaching a code of ethics or attempting to isolate underlying principles, so the medium of television with its continual 'flow' that resists the closure of judgement, keeps open the possibility of a variety of responses, some of which are shown, others implied. The viewers are in a position to consider for themselves which seem right and which wrong, often assisted by the responses of characters on the screen; the detective gathering information about the murderer, the revelation of how the surgeon treats his wife, the pundit's commentary on the footballer's moves. The viewer is not required or expected to deliberate on moral matters; all they need do is absorb the imagery of possibilities that the small screen contributes to the shared moral imaginary. Television does not socialise its viewers into the changing moral order through instruction but through showing new possibilities and engaging attention and interest. The viewer is seldom likely to respond by consciously taking up a new moral position as a result of what they have watched, but what is more likely is that ideas, values and attitudes are subconsciously taken in, perhaps on a trial basis, perhaps as an imagined possibility that may lead to later adoption or rejection.

Conclusions

Baudrillard claimed that the mass media 'are not producers of socialization but of exactly the opposite, the implosion of the social in the masses' (1994, p. 81). His radical belief was that the media are destructive of society and of meaning when they make their audience into a mass who receives the same message at the same time. His is an adaptation of McLuhan's reversal of the idea of the *explosion* brought about by mechanical technologies that extend the capacity of human bodies in space (through cars, cranes and tools), to the *implosion* created by the media because they bring everything to us where we are (1994, p. 3). McLuhan was right; television brings into our living rooms a version of society that does not replace but supplements and extends the social world we engage with directly. In a number of ways, we interact with what goes on through the small screen; sometimes we interact directly with people and places, at other times we are drawn into and included in interactions that are going on between fictional characters or 'real' people. What happens on the small screen becomes part of our everyday lives, echoing and emphasising the rhythms of the habits and routines that we usually follow. We become involved in the mundanity of other lives through the screen, some created, others just captured by the recording devices of camera and microphone. And it is not only the minute detail of fictional or real lives but also the more weighty issues of politics and the relations between institutions and states that we are encouraged to interact with.

Television extends the public sphere in two ways; firstly by including us in the reporting and accounting for events that have potential impact on many lives, and secondly by exploring in a public context the morals and mores of private lives. In the political arena, the public sphere needs minimal stage management to ensure that deliberation can follow the form of rational debate as closely as possible. But when private lives are exposed in the public sphere, stage management can enhance the poignancy and emotional impact of debates about how people have acted and how they should act.

The effect of extending sociation for viewers through the small screen is to keep them in touch, beyond their direct and personal experience, with a society whose moral culture is constantly changing. This is, I have argued, a form of socialisation for all the members of the society as they engage with ideas, arguments, values and practices shown on television that, while connected and related to their own, are nonetheless different. The formal institutions of socialisation (the family, the school, religion) are supplemented and extended by the mediation of the small screen directly into our homes. I have argued that the process of interaction through the small screen integrates into our everyday lives a continuous socialisation into the deliberations of the public sphere and the private sphere of the mores of others lives.

Baudrillard is, provocatively and interestingly, wrong when he suggests that the masses respond uniformly and as a silent majority in the era of electronic mass communication. The contents of the mass media – especially television – are not univocal in their content but produce a variety of messages that viewers can absorb or ignore, accept or reject, agree with or dispute. They may often be silent, but that does not mean that viewers have no response; they can take in what happens just as they do in other spheres of their lives. It would be impossible for viewers to hear a single moral message through a medium like the television because there are so many competing voices (channels, writers, producers, presenters) who each articulate different ideas and values and are keen to show possibilities rather than instruct viewers in a 'right' or 'righteous' moral path. Viewers are free to take what they wish from what they see on the small screen and do with it as they wish. And unlike Orwell's 'telescreen', they can switch off when they have had enough.

7
Mediating Morality

From showing to being moral

Up until now, I have argued that television simply *shows* us morality in its audiovisual content by presenting instances of good and bad human actions and their consequences. Programmes are usually designed and made to entertain, amuse, educate or stimulate viewers rather than to bring about a moral effect. I've argued that viewers' interest in the content of programmes has a moral edge, because just as people are interested in whether what those around them are doing is right or wrong, they have the same interest in the content of television. But television can also *be* moral in the sense that its own communicative action can have a direct moral effect; the audiovisual moving images are moral in *how* they show what they do. This is broadly the field of media ethics in which the viewers and those who intervene on their behalf, such as the state and regulators, critics and commentators, expect certain things of the media in return for taking what they see and hear seriously. The moral relationship is between the medium and the viewers and may be simplified in the idea that what is shown is *true*. Much of what appears on the small screen is fiction, and both those producing it and those receiving it are in no doubt that it is. Nonetheless, there is always presumed to be a measure of truth in what is shown, that the fictional represents the real truthfully in a general way while the specifics may not be literally true. Even jokes or comedy sketches are built on an exaggeration or distortion of the truth and depend for comic effect on a shared recognition of what is true that is being distorted. When what is being shown is presented as actuality, both producers and audiences know that it is an abstracted, summarised and truncated version of the truth and will always be lacking. Nonetheless, in all these instances, there is a presumption that the media are telling the truth and that its limitations are clear. The ethics of the media extend beyond simply expressing the truth because sometimes the truth should not be told, and sometimes how the truth is told must be managed so that it does not cause harm to individuals.

There are a number of different ways to approach this theme of television's moral responsibility as a medium. I will first look at the idea of media ethics and the moral panics that ensue when things go wrong. Then I will turn to the representation of distant suffering that has attracted considerable concern from social scientists in recent years and has opened up the theme of morality. Just how television operates as a medium that represents the world truthfully is nicely approached through the concept of 'witness' that has been developed from John Ellis's early work. And then I will turn to a tricky area of the critique of television that spells out its structural inability to tell the truth, one that we are complicit in, because of its role in postmodern society. Before developing these themes, however, it is worth dwelling a little more on the idea of 'truth'.

Just what counts as the truth varies according to circumstances as epistemologists and sociologists of knowledge have argued as they raise issues of conjecture, evidence, argument, correspondence, refutation, objectivity, perspective and relativity. But for pragmatic communicative purposes, the idea of the 'honest truth' works well; what is uttered, shown or written as being true has behind it the intention of the utterer, shower or writer to be truthful. This idea of truth is the opposite of lies, obfuscation, dissembling or knowing silence and is the sort of truth recognised in courts of law, the religious confessional and other institutionalised communicative contexts. The same presumption of truth lies behind ordinary conversational exchange; if what is uttered is not presumed to have the force of truth, then there is not much point in communication. From Aristotle to Žižek, the issue of truth has been recognised as a moral problem; social relations, from one-to-one interactions to those between a state and its people, depend on trust that what is said is true. And of course, when what is said in these various contexts is found to be lacking in truth, trust is withdrawn, and the relationship is put in jeopardy. The media are in a similar situation in which there is an expectation of truth although, as with politicians and advertisers, a weary recognition that the quality of truth is variable and unreliable.

Media ethics

Media ethics has two key principles that may often work together but can sometimes be in contradiction; one is to communicate the *truth*, the other is to communicate *appropriately*, that is, without doing harm. When these two issues are in tension, there is no formula, system or code of journalistic ethics that can resolve the dilemma, which is why Philip Patterson and Lee Wilkins argue that journalistic ethics must be taught by using case examples and philosophical ideas: '... thinking about ethics is a skill anyone can acquire' (2005, p. 3). They see the origins of journalistic ethics, not in philosophically based ethical systems, but as emerging in the early 20th century around the idea of *objectivity*, a perspective on the truth that resists any individual bias on what is reported (Patterson and Wilkins, 2005, p. 22). They go on to argue that even though this principle of objectivity was still valued at the end of the 20th century, it had been modified by a pragmatist view that the context of showing or writing the truth can affect what counts as true. The notion that truth is always context dependent has the potential to undermine the practices of good journalism – rigorous research, honesty, discipline, impartiality – that are oriented to speaking the truth. However, sticking to these practices balances the pragmatist claim that description always involves subjective evaluation against those objective characteristics of events that just cannot be ignored by good journalism (Kiernan, 1998b). There may, of course, be good grounds for withholding the truth because telling it may cause more harm than not doing so. Instead of simply telling the truth, journalists also have a responsibility to the subjects of their stories so that in presenting the truth they blend principles of accuracy, tenacity and sufficiency with those of dignity, reciprocity and equity. Recognising the dignity of those they are reporting about and respecting community values and diversity are all part of telling the truth in an appropriate manner (Patterson and Wilkins, 2005, p. 36).

Any notion of 'the truth' is particularly problematic in relation to photorealistic imagery, and Patterson and Wilkins present a number of cases in their textbook for discussion by journalists learning about ethics. For example, they consider a photograph in a newspaper of the desperate, but unsuccessful, attempts to save the life of a two-year-old child who was killed by her father and whose burnt body was visible in the background (Patterson and Wilkins, 2005, pp. 233–7). The journalists felt it was their duty to print the image, to represent truthfully what had happened, but they were overwhelmed by the volume of calls they received from an angry public who felt that it was inappropriate and unnecessary. Another example that Patterson and Wilkins (2005, pp. 226–7) report is how different media used the video, originally prepared and distributed as propaganda, of the beheading of the journalist Daniel Pearl. Pearl was abducted in Karachi, Pakistan, in January 2002, and a month later the FBI received a three-minute video that showed him admitting to being Jewish, apparently denouncing American foreign policy, and then, with his throat slit, being decapitated. An edited portion of the video was shown by CBS news in May 2002 as illustration of the power of the Internet to spread propaganda, and later that month the *Boston Phoenix*, posted a link to the unedited tape on its website and then published still photographs of the incident from the video. There were attempts by Pearl's family to persuade CBS not to use the video, and the Secretary of State's office and the FBI tried to persuade the video website to remove it. Showing images, especially moving images, of people who are dead or dying may be the truth and may show what terrorists are willing to do, but it is not necessarily appropriate. The Pearl video – which can still be found on the Internet – could be seen as undermining his dignity, challenging community values and playing into the hands of those who were seeking publicity for their cause. Precisely how such matters are to be judged depends on the values of a particular culture and will, of course, vary over time and across the world (see Chapter 3).

Television has the potential to shock its audience, and those countries that have long-established and independent television companies not controlled directly by the state often have a bureaucratic system for following up complaints from viewers. Different countries have different systems for licensing and controlling television companies, and regulatory authorities are often halfway between government and the makers of programmes. For example, in France, the *Conseil supérieur de l'audiovisuel* (CSA) has this role, and in Germany there is a more complex system with regional regulatory bodies for private television and internal structures for state broadcasters, while in Italy the Communications Guarantee Authority (*Autorità per le Garanzie nelle Comunicazioni* – AGCOM) brings together regulation and control but has powers that overlap with those of politicians at regional and national level. In the UK, Ofcom investigates possible transgressions of the Communications Act 2003 and the Broadcasting Code, which is periodically updated. Many of the transgressions it investigates are to do with product placement or the broadcasting of material before the 9pm 'watershed' that it is deemed inappropriate for children to watch. Sometimes complaints by individuals who feel aggrieved are investigated, such as the woman who had taken part in a programme about facial surgery on the understanding that her anonymity would be protected, but her first name was used, and enough of her face was shown for her to be recognised.

Occasionally, there are many more people moved to complain, such as the approximately 500 people who contacted Ofcom about remarks made by comedian Frankie Boyle in his comedy series *Tramadol Nights* broadcast on Channel 4 in December 2010. Boyle made 'jokes' about the relationship between a female celebrity, her ex-husband, her new partner and her son who has disabilities. The solicitors who acted on behalf of the celebrity summed up their case like this: '…the comments were discriminatory, offensive, demeaning and humiliating', and those for the other complainants said that it was: '…highly offensive, discriminatory and abusive to broadcast these comments about an eight-year-old disabled child. The complainants were also offended that the comments named a child as engaging in rape and incest'. What the Ofcom report also shows is that the broadcaster had the opportunity to respond and made the case that this was a: '…series which reflects Frankie Boyle's misanthropic brand of humour, in which he is both self-mocking and outwardly scabrous about the world at large'. Because the broadcaster had warned viewers about strong language and the potential that they might be offended, they 'reject[ed]…in the strongest terms that this [was] a joke about…disability, or about rape or incest – it [was] simply absurdist satire'. Ofcom didn't agree and decided that even absurdist humour could be offensive and that the material '…appeared to directly target and mock the mental and physical disabilities of a known eight-year-old child who had not himself chosen to be in the public eye. As such, Ofcom found that the comments had considerable potential to be highly offensive to the audience' (all above quotations from Ofcom, 2011b).

This is a case where the different sides disagreed quite strongly about the ethical standards that should apply; the broadcaster felt the truthfulness of Frankie Boyle's humour lay in his self-mocking misanthropy, but the complainants felt that they had been inappropriately harmed. The case shows how the broadcasting regulator follows a quasi-legal process that is deliberative, hears arguments from either side, weighs up the case and passes a judgement that has the force of law. Ofcom judged that the broadcaster (not the comedian) should have anticipated that the offence would be of a degree sufficient to constitute harm and should have edited out the material, effectively censoring it.

Philosophers who comment on the issues of media ethics recognise that attempts to control harm through regulation or law is reasonable, but actual harm done by television or any other medium can seldom be identified. Damage to a person's reputation through, for example, implicating them in a sexual affair or financial swindle, can be a consequence of trying to tell the truth, but any harm caused must be shown to be sufficiently great to counterbalance public interest in the truth (Graham, 1998). To persuasively argue that privacy has been invaded, for example, the complainant needs to show that there was an inappropriate attempt to get at the truth rather than that the media were simply making public what some people already knew (Archard, 1998). In the UK, the Press Complaints Commission is an industry-sponsored organisation with voluntary cooperation and no legal powers, whereas the broadcast media are much more tightly controlled with Ofcom as a statutory organisation guided by a code and the law. In the case of *Tramadol Nights*, Ofcom was not moved by concern for the celebrity's privacy so much as the offence caused to the public and the potential harm to the child in being associated with the offensive 'jokes'. Anthony Ellis (1998) argues that censorship on grounds of possible harm is very difficult to justify, and the potential for offence cannot be sufficient grounds for censorship – although collective outrage might just be sufficient threat to public order to warrant restraint. That is why broadcasters like Channel 4 are careful to warn viewers of programmes like *Tramadol Nights* that they might be offended, because then it becomes the viewers' responsibility to protect themselves and their children from what might lead to outrage.

These few examples give enough of a flavour of what is a complex area of professional ethics to demonstrate that what is shown on television – particularly broadcast television – may be judged by regulators and others to cause moral harm. The sorts of harms are offence to individuals through lack of respect, through untruth (even when it is meant to be a joke), through invading privacy or through causing offence to the public, that may lead to outrage. It is important that those who make television programmes consider possible harms, just as it is important that they follow practices intended to ensure they are telling the truth. The existence of a regulator provides the public with an institution through which to exercise some restraint on what is broadcast – but by and large, the moral import of the issues it deals with is slight. There is a risk that regulative control on television becomes a form of censorship that could be used to protect the establishment, and in particular the state, which might very well be offended if their impartiality or right to power is opened to ridicule. Freedom of speech is an important principle of democracy that professional media ethics needs to uphold. The media, including the audiovisual media, are keen to keep their audience, and this is usually sufficient to deter them from risking offending even small numbers of viewers. A much greater risk is that the regulators make broadcasters too fearful of being punished so that they impose a form of self-censorship, which would then reduce the capacity of television to contribute to the public sphere. Interference from the state or from other major social institutions is the greatest threat to the moral diversity that the small screen contributes to the moral imaginary.

Moral panics

Failing to follow professional ethical standards may lead to what is shown on the small screen causing harm or outraging public morals. Another way that the small screen can have a direct moral effect is through contributing to a moral panic. Moral panics arise when the behaviour of a group of people is identified in a stereotypical and stylised way in the media as being a threat to established social values or interests. The media cannot create a moral panic on their own, they need the contribution of 'moral entrepreneurs', groups or organisations 'who take it upon themselves to pronounce upon the nature of the problem and its best remedies' (Critcher, 2003, p. 17). Moral entrepreneurs are the politicians and pressure group spokespeople who often claim to represent public opinion and who express outrage in a way that amplifies events reported in the media. The media often make the distinction between 'experts' with a professional interest in the event, and those with a political or campaigning interest. One media outlet alone cannot make a moral panic; it requires repeated reporting of the issue and concerns in different parts of the media, whether it is the spread of AIDS, raves, video nasties, paedophilia or child abuse. Although the tabloid papers, with their sensationalist reporting, are usually at the forefront of moral panics, television is very likely to become involved by reporting the news, commissioning documentaries and even developing fictionalised storylines. Television cannot afford to ignore issues of public interest.

There is a recurrent concern amongst some media commentators and pressure groups that the use of bad language and the display of sexual activity and violence on television lead not only to offence and outrage but also potential harm, especially to children. Periodically, this becomes a moral panic, and moral entrepreneurs, such as the US pressure group *Parents Television Council* (PTC), organise campaigns to get the Federal Communications Commission (FCC) and other regulatory bodies to tighten the restraints on what is shown on television, particularly when children are likely to be watching. In the UK, *Mediawatch*, formerly the *National Viewers and Listeners Association* (NVLA), campaigns against 'violent, sexually explicit and obscene material in the media' that it sees as harmful or offensive.[1] In general, such pressure groups are an expression of the moral values and tastes of a section of society and work hard to commission and promote research that suggests their views are shared more widely. But their activities can lead to something like a moral panic when they draw politicians and academic experts into the debate and attract the attention of the media. The media generally wish to be seen to respond if there is any evidence of harm, but they will also respond if it simply appears to be a widely held matter of taste.

The NVLA, with its media-savvy leader Mrs Whitehouse, began campaigning in the 1960s in response to what it identified as a change in the content of broadcasting. As Tony McEnery (2006) points out, television ownership in the UK had increased rapidly in the post-war period to 90 per cent by 1964, and broadcasting began to present a broader, less middle-class view of society. The realism of some early BBC programmes such as *Play for Today*, *The Wednesday Play*, *Till Death Us Do Part* and the documentary series *Man Alive* attempted to show working-class lives as they were lived, including the bad language, violence and sexual activity. This televised material generated a response from Mrs Whitehouse and the NVLA, who developed what McEnery calls a 'moral rhetoric' to amplify the offence of the programmes they objected to as 'filthy', 'revolting', 'brutal', 'irresponsible', 'weak' and 'degrading' (2006, p. 9). McEnery's interest as a linguist is in bad language, but he points to the 'discourses of panic' in the way pressure groups such as NVLA expressed their concern over the moral threat about what appears on the small screen (for a cultural historical perspective see Smith, 1985). Such discourses tend to bundle together issues (for example, bad language, sexual and violent behaviour) as a single category, even when there is no evidence that they are linked. The category is then linked to a scapegoat, the television broadcaster, who is responsible for moral decline through a 'moral relational device' (Housley and Fitzgerald, 2009) that establishes blameworthiness through terms such as 'pornography', 'obscenity', 'permissive' and 'four-letter'.

Although Mrs Whitehouse's campaign to 'clean-up' television largely failed to alter the trend towards using bad language and showing scenes depicting explicit sex and violence, the threat of a moral panic continues to constrain makers of programmes. Rather than risk outrage and extensive offence, which may lose them viewers, programme makers work collectively in companies and corporations, to develop a sense of the boundaries of what will be acceptable. Nonetheless, the last 40 years have seen a gradual shift in what is acceptable in terms of the representation of nudity, sexual activity, death and violence, both in fictional and in actuality programming. What is offensive enough to cause outrage on television has altered and the only significant concern that remains is of possible harm to young children who might be expected to be watching before 9pm. Interestingly, a different mode of moral discourse has emerged that is concerned not with constraining broadcast television but with ensuring that it remains free of constraints from undue influence by commercial, political and sectarian influences. Formed in 1983, the *Voice of The Listener and Viewer* (VLV) takes a different line, seeking to 'safeguard the quality, diversity and editorial integrity of broadcast programmes' and 'promote a wider choice of high quality programmes'.[2]

The idea of a moral panic suggests that the outrage is excessive in relation to the actual or even possible harm that might result from the media. What is sad is that there is often a problem or behaviour behind the moral panic, which is of real concern – the spread of AIDS, child abuse, paedophilia or sexual violence, for example – that does require an institutional response and the alerting of moral culture. But what the moral panic does is attribute particular values to particular actions and actors in the abstract, and at the level of description connect the two to exaggerate consequences and causal agents through linguistic 'moral relational devices'. This type of discourse tends to bundle together both actions and actors into categories that are not subtle in discriminating evidence of actual causal relationships. The moral rhetoric used attaches values to phenomena that are not experienced directly but perceived only at a distance. The media literally *mediate* the actions and activities of other people who are, by definition, not co-present to viewers and readers. The audiovisual medium of the small screen promotes an illusion of co-presence and involvement and this gives apparent substance to the categories of people described. This is why it is so powerful in reinforcing discursive attributions of cause and helps to make the actions of particular agents appear to be more dangerous than they actually are.

The depiction of violence or sexual activity on the small screen, whether simulated or actual, can lead to a sense of involvement and obligation to act in relation to what is shown, as if one were there. Watching actors on screen pretending to make love puts the viewer in the position of the unintentional voyeur; curious but embarrassed. Seeing violence towards a person can lead to shock, fear, indignation and the desire to intervene on behalf of the victim. These are responses that are both visceral and moral; our feelings are aroused, but we also know how we *should* feel confronted with these things. When we are confident that what we are seeing is simulated, that it is not real, we can retreat into the aesthetic response of pleasure in arousal, enjoying the playing out of the situation, perhaps hoping that things will work out well for those to whom we feel sympathetic. The moral campaigners are concerned that our engagement with these unreal experiences will desensitise us to appropriate moral responses to sexuality and violence in the world around us – but they have yet to establish persuasive evidence that anything like this happens.

Distant suffering

The audiovisual moving image can bring distant places and the very different everyday lives and experiences of the people in them, through the small screen to the television viewer. Viewers can see the mores –, the ways of doing and acting – of distant people but the most poignant visions of other lives are of those who are suffering; such imagery reveals the 'moral consequentiality of television', as Tester (1999) puts it. The potency of audiovisual media to communicate emotionally means that moving images of the bodies of starving children, accompanied by their cries, have more impact than the words 'starving children' in a written report. It is difficult not to be moved by such images, which seem to need little explanation other than perhaps answers to the questions 'Why has this happened' and 'What can I do?' Michael Buerk's report from Ethiopia in 1984, accompanying the work of cameraman Mohammed Amin, showed the starving people who had gathered in a camp near Korem where Save the Children had set up a field hospital. It brought home, literally, to those in the UK the plight of people, especially children, who were dying three and a half thousand miles away.[3] What is striking about the imagery is how different the experience of the starving people is from that of the people watching television; it is difficult to relate to their situation, to imagine it happening at home, to understand the extent of the human suffering and degradation through poverty and malnutrition. The 'truth' of the situation – these are human beings, who are actually in this situation – is manifest in the audiovisual imagery as the camera scans across people sitting and lying in the camp, zooming in on faces as some move listlessly about. We see mothers comforting painfully thin, crying children, and we see the wrapped bodies of small children who have died being carried from the camp, the faces of their parents contorted in grief. The commentary gives us facts – 7,000 babies being cared for by Save the Children Fund, another 114 babies have arrived this morning, 40,000 people in the camp – as the camera moves along the queues of families patiently waiting for help, stopping to capture the shocking image of flies crawling undisturbed on a live person's face. The remarkable power of television to bring closer the lives of people who are distant from the viewer, not only in miles but also in terms of cultural and life experience, has attracted the interest of a number of social scientists and commentators (Ignatieff, 1985, 1998; Boltanski, 1999; Tester, 1999; 2001; Chouliaraki, 2006).

The televising of distant suffering in this way led to a very televisual response with *Live Aid,* a television spectacular of pop and rock music in 1985 that was explicitly linked to raising money to help those in the Horn of Africa. The desire of viewers to do something, to express their collective concern, led not to a political expression but a media event (Dayan and Katz, 1992 – see Chapter 3) that brought viewers together across the western industrialised world. The spectacular owed much to the pop festival – a series of bands playing outdoors to a large number of people over many hours – but was oriented to the television audience that has been estimated at over 1.9 billion in 150 countries. The event took the form of a concert with simultaneous performances at Wembley stadium in the UK and the John F. Kennedy stadium in Philadelphia. The musicians – many long-established major stars of rock and pop music, including Paul McCartney, Mick Jagger, David Bowie, The Who, U2, Queen, Madonna, Joan Baez, Tom Petty, Bob Dylan, Tina Turner, Crosby Stills and Nash, and Black Sabbath – apparently performed for free, while those attending the concerts paid an entry price, and those watching at home on television were asked to make donations.

The televised event apparently raised something like £150 million, and although there were questions about how much of the money got through to actually help those in need, it set a model for televising charitable money raising; celebrities in the west give special performances, and the television audience contributes money. A modified version of this involves audience members taking part in sponsored activities to raise money, which is then contributed through the nexus of a television programme to a 'good cause'. The celebrity performance and the active audience are often combined as through the *Comic Relief* events that also began in 1985, inspired by the starvation in Ethiopia. Television marathons, or 'telethons', charity fundraising events that involve programmes, or a series of connected programmes lasting a number of hours, have a longer tradition in the United States (see Tester 2001).

The suffering of strangers, seen through the television screen, seems to demand a response that is not systematic or rational but is emotionally cathartic for the viewers who can get excited and enthralled by the spectacle, the celebrities and the music or jokes.

The televisual response to the suffering of strangers suggests a coming-together of viewers, a sharing of emotion and sympathy that involves a celebration of that sharing, of the collective conscience. The co-present audience at the spectacular and the 'ordinary people' in the telethon are agents for the viewers at home, representing their emotional concerns. The rational outcome may appear to be money; it seems to be the only way that the television audience can connect and communicate with the suffering of strangers. But these programmes also appear to engage viewers morally in the lives of distant strangers, other people who are not part of their community. This is a form of what Michael Ignatieff (1998) calls 'moral universalism' – a sense of moral obligation to those who are recognised as fellow humans even though their social and cultural situation is very different.[4] Ignatieff expressed a disquieting ambivalence when he wrote: 'The medium's gaze is brief, intense and promiscuous...' and went on to spell out the moral distance that appears to be traversed by the immediacy of live, moving (in both senses) imagery that makes the viewers into voyeurs (1985, p. 58). As he traces the philosophical and Christian history behind bourgeois moral universalism, he notes that there was always a hierarchy of obligation that put the stranger behind those known and nearby. The visual imagery may stimulate moral empathy in the viewer, but it obscures the historical and political relations that maintain the social distance between television watcher and famine victim. There is a risk, as Keith Tester (2001) argues, that moral universalism is seen as natural, part of the human condition, whereas it is actually a cultural phenomenon created by societies. Empathy with distant suffering does not seem to conflict with the international relations that, through trade, loans and aid have contributed to the political and economic circumstances of famine – although the Live 8 concerts, televised spectacular music events held 20 years after Live Aid, were an attempt not only to raise money but to draw attention to the policies of the G8 countries and their effect on world poverty.

But more important from Tester's viewpoint is the failure of television to continue to move its audience who become bored and apathetic as they suffer from 'compassion fatigue' through the repetitive use of the same sorts of stories and images (Tester, 2001, pp. 13–42; Chouliaraki, 2006, pp. 112–4). People may also become apathetic about other people's suffering when it is co-present and not mediated by television. Those who live and work amongst pain and suffering must become sufficiently inured to the initial impact of exposure to it to be able to continue their work. Those who perpetrate it as soldiers, terrorists or guards must develop a rationalisation that makes causing pain and suffering bearable. But because the proximity of distant suffering on the small screen is always through a quasi-interaction, rather than an embodied interaction, the non-engagement of switching off, whether of attention or the set, is always possible. It was noticeable in 2011 during another famine in the Horn of Africa with equally terrible consequences that the suffering was not presented on television in the same way as in earlier years. There was no televisual spectacular, no special event that has engaged the population, just repeated calls for donations and bleak news reports of the numbers of children dying.

Seeing people through the small screen who are suffering raises the problem of proximity; if it were happening to those close to us, what would we do? And how close is close enough; family member, neighbour, friend, community member, a member of the same society or member of the same species? Luc Boltanski (1999) points out that an obligation exists in those routine relationships of family, friends, neighbours and community because there is the possibility of a social sanction. At the very least, others will show disapproval if the response is not appropriate. But where there is no prior social relationship, when those involved are strangers, there are a number of possible responses to the suffering of others; ignore it, speak out against it or sacrifice something for it. The Live Aid and Live 8 concerts were attempts to speak out and sacrifice money or economic gain.

The classic dilemma of how to respond to the suffering of a stranger is told in the parable of the Good Samaritan reported in the Gospels, in which a Jewish traveller is left, beaten and robbed, lying by the roadside. A priest and a Levite both pass by on the other side of the road, ignoring the wounded man, even though they come physically close to him, while the Samaritan sacrifices his time and effort and perhaps his sense of ethnic difference. The story plays on both spatial proximity – the wounded man is bodily present to those who pass by – and cultural proximity because the Samaritan who does help the robbed man comes from an ethnic group who were hated by Jews, and Jesus tells it in answer to the question 'Who is my neighbour?'

A philosophical view, such as that offered by Raimond Gaita (1991), opens up the question of how the suffering person is responded to; as another animal, or as a fellow human identifiable not through cultural equivalence but through having a 'soul', that is, the capacity to feel in a human way. Gaita's notion of soul comes not from a religious context but his reading of Wittgenstein and he comes to the conclusion that the response of pity is founded on our reaction that distinguishes human suffering from that of any other animal '…because of the meaning pain can have in a human life and because of what a human life can mean' (Gaita, 2004, p. 186). Gaita's inquiry into what counts as 'good' goes rather further than the Good Samaritan, and while he is interested in the cultural dimensions of both goodness and humanity, he does not discuss awareness of suffering through the small screen. His discussion of the parable does, however, tell us that moral understanding is not only about what people do but also what things *mean,* and this helps us to see that what makes the audiovisual image of suffering so powerful is that the viewer cannot deny the meaning of the human suffering they are watching. The apperception of suffering cannot be dismissed as mere description, words that may be true or not, an account that can be interpreted in one way or another, but must either be confronted with sympathy or with eyes that do not see the suffering.

Boltanski argues that distance, the lack of physical or social proximity, negates personal obligation, and a politics of pity comes into play as the viewer becomes a spectator of events playing out on a world stage:

> What particularly holds our attention in the character of the spectator is, on the one hand, the possibility of seeing everything; that is of a totalising perspective of a gaze which has no single point of view or which passes through every possible point of view and, on the other, the possibility of seeing without being seen. (Boltanski, 1999, p. 25)

While a politics of justice addresses who is to blame, the politics of pity regards individuals as suffering *en masse* and does not question whether they deserve their fate; their situation is taken to be bad luck. For Boltanski, the images of suffering are not enough to create a moral relationship between spectator and victim; the commentator, such as Michael Buerk in Ethiopia, is necessary to allow the viewer to fill the role of the 'pure spectator'. The agency of the reporter must convince us that what we are seeing is true, reasonable and justifiable, but their agency is also needed to turn the viewer into a 'moral spectator' who is able, through an act of imagination, to put themselves into the situation of those suffering. Boltanski's 'moral spectator' is derived from Adam Smith's 'impartial spectator' who, as well as those suffering, is able to observe themselves and have a sense of how they should be seen reacting to the spectacle of suffering (see Chapter 8, pp. 199–205). The moral spectator is not simply reacting with emotion but acknowledging a social and cultural responsibility that recognises what it *means* to be suffering and what it *means* to recognise it.

The presentation of distant suffering on television can, of course, be presented in a variety of ways, and the use of spoken and written words and the devices of televisuality will alter the impact it has on viewers. Lilie Chouliaraki (2006) has played close attention to how mediation creates different spectator positions through analysing broadcasts of different types of distant suffering. She argues that there are different ways of representing the suffering of people at a distance that invite different modes of spectatorship established through the '*socio-cultural* immediacy' of television that can '...facilitate *modal* imagination. Modal imagination is the ability of spectators to imagine something that they have not experienced themselves as being possible for others to experience. It is not simply impossible or unreal' (2006, p. 20 – emphasis in original). It is through modal imagination that the distance of those who are suffering can be reduced, and the viewer's sense of connection and involvement enhanced. For Chouliaraki, there is an issue not only of how the media present distant suffering but also of the public context of engaging in a politics of pity that can create solidarity between viewers and sufferers. The way that the media show issues modulates this relationship around what Silverstone (2007, p. 49) calls 'proper distance'. He argues that it can be too close or too far, too intrusive or too dismissive, and that mediated distance is an achievement of both the makers and the spectators of the media: 'Both actors and spectators are present in the public space, both actors and spectators must carry the burden of judgement' (Silverstone, 2007, p. 48). Chouliaraki (2011, p. 376) warns that the 'agonistic solidarity' of proper distance is threatened by those televisual devices introduced to counteract the effect of compassion fatigue; things like branding, celebrity endorsements and graphic animations that get between the voices of the suffering and those who wish to help. Attempts to bridge distance with sentimentality, seduction or voyeurism undermine the politics of pity, she suggests. Another way of putting this is to suggest that the realism or truth of what is shown must be appropriate and honest.

One of the most disturbing elements of Luc Boltanski's subtle reflection on distant suffering remains the potential for an aesthetics of pity to intrude into the politics of pity. Viewers may be moved to denounce the actions of others, such as the Ethiopian government's policies during the famines of 1984–5, or they may be moved to sympathy and giving money through aid agencies. However, the aesthetic response of gazing in fascination, finding beauty in destruction and suffering, excitement at the emotions aroused by the images of suffering, is the response to suffering that maintains most distance. The aesthetic response is, of course, cultivated by much television and movie drama in which disaster of some sort is the theme of the unfolding action, and the pleasures of watching are in the feelings of shock and horror, the indulgence in the fantasy of how life could go badly wrong and the thrill of being moved to imagine what it might be like without risking having to experience it. The same sort of aesthetic response can be provoked by so-called 'reality' television in which unpleasant experiences, illnesses, disabilities or situations are the topic. The televisual spectator is able to put her or himself in the shoes of the other person but enjoy the fact that they are not actually in those shoes and can return at will – or at the end of the programme – to their own much more comfortable shoes.

But is it real?

It is embarrassing to contemplate the idea that the nobility of the destitute, the beauty of poverty and the courage of those who are disfigured or disabled can be aesthetically fascinating, yet generations of art lovers have been draw in by the suffering of others depicted in paintings such as Théodore Géricault's *The Raft of the Medusa*. This exemplar of French Romantic painting, first shown in the Paris Salon in 1819, depicted a real shipwreck in which only 15 of 147 people on the raft survived. The composition of the picture is classical in its aesthetic form, designed to attract and guide the viewer's eye, but it was also graphically realist in its representation of waves and clouds, wood, rope, cloth and bodies, both alive and dead. The predominantly realist mode of television is photographic rather than simply graphic, and the effect is that viewers presume that what they see is the truth. If what was shown as actuality was not presumed to be 'true', then ethical standards would not be so important, television could not contribute to moral panics, and the showing of distant suffering would have little impact.

The realism of television is, I have argued, what enables viewers to make sense of it with the same resources as they use to make sense of the everyday world around them. This is, however, a rather unfashionable view because many commentators have been concerned to point to the social construction of the televisual 'message'. Theodor Adorno (1991 [1951]) was an early critic of television as a cultural form, who identified its multiple or 'polymorphic' layers of meaning superimposed on one another that obscure a 'hidden message' and work like 'psychoanalysis in reverse' (a phrase he attributes to Leo Lowenthal). As yet another instance of industrialised culture, he also pointed out how television deals in stereotypes and engages in 'pseudo-personalisation'. In a similar vein, John Fiske warns against precisely the window/mirror metaphor of the small screen that I have used in Chapter 5, as he explains how realism is 'defined by what it says the real consists of' and in so doing it achieves 'the positioning of the reading subject through its form' (Fiske, 1987, pp. 24–5). Fiske's deep suspicion of the capacity of television to present the world as it is, treats realism as an ideological device and he describes the televisual 'metadiscourse' that makes use of lighting, camera angles, close-ups and the script to create a 'reality' that is persuasive but illusory. Having established realism as hollow, Fiske becomes more interested in the 'active audience' and the potency of diverse readings that can lead to the 'power to construct meanings, pleasures and social identities that *differ* from those proposed by the structures of domination' (1987, p. 317). Strangely, though, he recognises the potential for a simple realism in 'cheaply produced dramas and soap operas' in which the technical capacity to construct reality fails with the consequence that:

> The absence of authorial (or editorial) intervention adds subtly to the sense of realisticness, the sense that the camera is merely recording what happened, and to the sense of liveness, that it is happening now – the same perfect match between represented television time and the "lived" real time of the viewer is, after all, characteristic of genuinely live television, such as sport. (Fiske, 1987, pp. 22–3)[5]

'Perceptual realism' is the term Noël Carroll (1998) uses to refer to the fact that the 'recognitional capacities' we use to navigate the television image are 'are the same or pretty much the same as the ones that we use to navigate everyday life' (1998, p. 138 – see Chapter 4, p. 106, above).

Just because what appears on the screen appears real does not mean that audiences should simply accept the reality as it is presented. Critical attention to what is shown has raised important issues about how society is represented – the distortion of class, race and gender, for example – and this has relevance not only for politics but for also for the moral culture. Some makers of television programmes have responded to this tradition of criticism by making programmes that attempt to shift or subvert stereotypes and pressure groups. For example, Channel

4's *Cast Offs* (originally broadcast in, 2009) used six actors with identified disabilities (dwarfism, cherubism, paraplegia, deafness, blindness and thalidomide disability) and a cast that included some non-disabled actors in a drama that pretended to be a reality show in which the six key characters had been left to fend for themselves on an island. Scenes of the characters dealing with situations on the island were interspaced with them reflecting on their experience after making the show in a way that played with disability both through perceptual realism and through irony.

For example, the actor with dwarfism playing 'Carrie' in the first programme in series one, is shown walking along and saying that she learnt a lot about disability on the castaway island although 'you know, I felt a bit bad, not being disabled', then when she comes to a call system for a gated block of flats, she cannot reach the buttons. Explaining what she means by 'disability', she gives the example of Dan, who was in a wheelchair, 'that's very different to me, I'm just a bit on the short side', before getting someone to open the gate for her. This realist demonstration of the disabling effect of the presumption of 'normal' height in building practices is in contrast to the irony of her scripted talk. It is also in contrast to the acted-out heavy irony of the (non-disabled) person she is visiting, who feels embarrassed about his height relative to hers and tries to get down to her level as they talk. First, he bends his knees, leaning on an armchair as he gets lower, then he sits on a sofa and slouches lower and lower, trying to maintain level eye contact with Carrie...but she then leans on the arm of the sofa and begins to lower her face in relation to his, forcing him to slide even lower as she bends her knees. By playing with perceptual realism, the makers of the show provoke an awareness of disability that would be difficult to achieve discursively.

The use of irony to blunt perceptual realism lies at the heart of Jean Baudrillard's provocative series of articles republished as *The Gulf War did not take place* (1995), about the media reporting of the 'war' as it built up, took place, and its aftermath. Many commentators were angered by Baudrillard's insensitivity to those whose lives were lost or damaged during the war, but they were ignoring the irony, just as anyone who accused Kiruna Stamell's delivery of 'Carrie's' script in *Cast Offs* as disablist would be. Baudrillard was pointing to the managing of the information about the war, the way in which its progress was planned and devised to meet the needs of the media such that what we were shown on television was a simulacrum; however perceptually real the images on the screen may appear, the audience could have little faith that they represented the war as it was experienced by those involved.

Before the war took place, the United States prepared for it with simulated video wargames, and Baudrillard pointed out that it would not be like previous wars because of increasingly skilled manipulation of the media: 'War has not escaped this virtualisation which is like a surgical operation, the aim of which is to present a face-lifted war, the cosmetically treated spectre of its death, and its even more deceptive televisual subterfuge...' (1991, p. 28).

The metaphor of the surgical operation proved to be prophetic of the media war that delighted in the missile's-eye-view images as they exploded on their targets – but the television did not show the images from those that missed their targets or hit civilians. What the television did show were dramatic images of US planes, tanks, missiles, ships and complex sophisticated equipment that could dispense death and destruction. There was also footage of troops and of generals telling viewers about their enthusiasm for the war, of their sense of justice and of the duplicity of the Iraqis who were locating their weapons near civilians. There were high-contrast night shots of tracers, of fires burning or missiles taking off and landing with dramatic explosions, and there were images of Iraqi soldiers surrendering and their burnt-out tanks. But there were very few images of the disruption to lives of people who were not fighting, of the victims or survivors of western attacks, of the injury and death that was caused. The images that were shown were those that the military forces were happy to see released, but as Baudrillard pointed out:

> It is a masquerade of information: branded faces delivered over to the prostitution of the image, the image of an unintelligible distress. No images of the field of battle, but images of masks, of blind or defeated faces, images of falsification. It is not war taking place over there but the disfiguration of the world. (Baudrillard 1991, p. 40)

Perhaps the worst that can be said of Baudrillard's critique is that it only tells us what we already know; perceptual realism does not necessarily mean it is the truth. Even the earnest attempt to tell us what is real, through serious news reporting, is confounded by the military and political need for acceptable propaganda once war is announced. This does not mean that viewers take what they see on television at face value because they understand that journalists' access to war zones is restricted by the military, if only to maintain security and control the flow of information to the enemy. Ironically, it was the complaints from journalists of lack of access to the Gulf war that led to the American practice of 'embedding' journalists with military units during the invasions of Afghanistan and Iraq. Although this practice generates more realistic images of the battlefield, the journalists become dependent on, as well as emotionally attached to, their soldier colleagues. As Baudrillard put it: 'Information has a profound function of deception. It matters little what it "informs" us about, its "coverage" of events matters little since it is precisely no more than a cover: its purpose is to produce consensus by flat encephalogram' (1991, p. 68).

The destruction of the twin towers of the World Trade Center (WTC) in 2001 was a mediated event, televised live as well as constantly repeated, which had a colossal symbolic impact that is still reverberating. The realism of the television images of the planes hitting the towers, of the fires and the collapse of the buildings, of people jumping, of people appalled and running, of firemen and policemen courageous but helpless, undermined perceptual realism. What was shown on news broadcasts around the world looked more like the fakery of a disaster movie than anything that could be treated as real; it was only the emotion-laden accounts of those who were there that lent reality to the images. Baudrillard, writing less than a year after the event, recognised its symbolic potency as fundamentally moral: 'Terrorism is immoral. The World Trade Center event, that symbolic challenge, is immoral, and it is a response to a globalization that is itself immoral.' (2002, p. 12). The often-seen moving images of the end of the WTC are an example of how the realism of the imagery did not in any simple sense make what was on the screen real.

I remember seeing the images shown on a screen in a university coffee bar, first as I made my way to get my coffee and then on my way back to my office. What I saw over the heads of people standing, transfixed, hands up to their faces, clearly made sense as moving imagery. The clips were being repeated, and excited commentators were trying to grasp that these were real images and not faked, that this was no practical joke or public relations stunt. The horror of what we were seeing on the screen was nonetheless distant; distant in terms of miles (I had walked beneath the WTC towers a few years earlier, but it still seemed a distant and exotic place) and distant in terms of culture because the excessive, explosive and expansive nature of the images seemed very American. What to do with it? How to make sense of it? How to connect with it?

From real to Real

Also writing within a year of the event, Slavoj Žižek articulates this problem with encountering the real through a medium that is constituted by images that contribute to the stock of our imaginary lives: '...the question we should have asked ourselves as we stared at the TV screens on September 11 is simply: *Where have we already seen the same thing over and over again?*' (Žižek, 2002, p. 17 – emphasis in original). A couple of pages earlier, he had spelt out the answer: '...for us, corrupted by Hollywood, the landscape and the shots of the collapsing towers could not but be reminiscent of the most breathtaking scenes in big catastrophe productions' (Žižek, 2002, p. 15). Those Hollywood disaster movies are reflected in a small way in many 'cops and docs' dramas on television and every local disaster that disturbs the smooth running of life in a soap opera. They serve a similar role in allowing viewers to see their nightmares depicted on the screen; the suspension of the paramount reality of their lives in which they can take for granted the presence of others and the functioning of material life. It is the possibility of discovering that the reality we experience is not as we thought it was that is fascinating to us as viewers.

The interest in anticipating and depicting reality through the media during the 20th century as carefully as possible, Žižek calls the 'passion for the Real'. It has led to the reconstruction and creation of fake experiences of death and disaster for on-screen drama but also helps us to engage with the real disaster of earthquakes, wars, tsunamis and famines. The possibility of the small screen is that we can be taken there – to the site of the disaster, to an imagined horror, to a reconstruction of an earlier experience of the boundary of humanity – while still in the safety of our homes or coffee shops. The Lacanian interpretation of the process that Žižek gives is that there is a tension between the real – reality as we perceive it and experience it in our own lives – and the Real – the reality that we have a sense of that exists beyond our capacity to represent it in language, beyond the imaginary, beyond reduction to symbolic representation. The experience of extreme violence, even on the screen according to Žižek, is experienced in a dreamlike way that is difficult to match with the reality of ordinary experience, and the gap means that:

> ...when, afterwards, we return to our 'normal' reality, we cannot conceive of both domains as belonging to the same reality. The reimmersion in 'ordinary' reality renders the traumatic memory of the horror somewhat hallucinatory, derealising it. This is what Lacan is aiming at in his distinction between reality and the Real: we cannot ever acquire a complete, all-encompassing, sense of reality – some part of it must be affected by the 'loss of reality', deprived of the

character of 'true reality', and this fictionalised element is precisely the traumatic Real. (Žižek, 2001, p. 66)

Paul Taylor (2010, p. 66) explains that the Lacanian Real is in contrast with the conventional, common-sense notions of reality. Something extra is needed to intervene to shock the viewer out of their immediate reality (concerned with getting the coffee back to the office...) to feel for real the suffering they have seen. The images of the destruction of the WTC are easiest to make sense of as a hallucination, to 'derealise' it by wondering how it might have been faked. The capacity of the audiovisual representation to provoke the imaginary while resisting reduction to the symbolic (no amount of words spoken could make the images make sense), has the potential to interrupt the conscious experience of paramount reality like a waking dream, generated from outside our own psyche. We can only experience reality as it is for us directly, but we can get a glimpse into the reality beyond as experienced by other people through the medium of the small screen. Whereas reading requires the reconstruction of the reality of others' experience through images developed within each reader's mind, the camera and the screen can capture, construct or reconstruct their reality in a way that is apprehended as a dynamic audiovisual image by all who watch.

The collective realm of other people's experience structured as the 'Other' is part of Žižek's Lacanian toolset, as Taylor explains: 'The big Other is the Lacanian term used to describe this intangible structure, the social repository of collected and projected beliefs, which we all relate to and rely upon' (2010, p. 73).[6] To be a social subject, one needs to engage with how other people deal with the world and the audiovisual media of the small screen help to constitute the big Other for us. The need for the repository of other people's experience that is the Real to connect with our own lives means that there is an ever-increasing push for the media to be real, or at least realistic. For what is shown to be accepted, it has to be seen as 'relevant', 'convincing', 'likely' or 'probable' and these sorts of judgements would be made about science fiction or murder mysteries as well as actuality programmes, albeit that viewers will grant a different sort of narrative licence to these different genres. The increasing technical capacity for the audiovisual media to reproduce reality seems to keep pace exactly with its capacity to fake or create reality. Taylor (2010, p. 74) puns on the so-called reality television show *Big Brother* / Big (Br)Other as an instance of the small screen trying to compensate for the inaccessibility of the Real by allowing the real lives of people to play out on the screen while the viewers watch.

The contestants are, however, inside a carefully controlled environment that is cut off from the real world – somewhere between a holiday camp and a concentration camp – as most starkly evidenced by their lack of access to media or communications other than with each other or us.[7]

Zygmunt Bauman has pointed out that as a game of exclusion, this makes for a particular morality: 'Excluding others instead of being excluded oneself (that is, excluding others before your turn comes to be excluded) is the name of success' (Bauman, 2002, p. 65). The affection and care that contestants develop for each other, the trust and fellow feeling as well as the simple pleasure in their company is constantly undermined by the threat of loss. And yet, the loss of a friend when they are voted off means that you have not lost the game, so each contestant is always ambivalent about whether another stays. If the contestants can engage in human interaction under these strange circumstances, the viewers are kept at a distance, not having any real relationship with them other than through voting.

Strangely, Bauman thinks that 'the spectacle of Big Brother bears an uncanny resemblance to the all-too-familiar experience of the spectators' (2002, p. 62). For Bauman, this and other shows such as *Weakest Link* (BBC One) demonstrate the 'disposability' of humans in a game of life in which each player must look out only for themselves because: 'If you are not tougher and less scrupulous than all the others, you will be done by them – swiftly and without remorse' (2002, p. 63). It is not clear whether Bauman sees the reality shows as reflections of how life is in late modernity ('This is how the viewers felt they lived all along, but now they see it vividly and clearly ... ') or whether he fears that they are treated as models of how life should be lived by their spectators (the shows ' ... tell their viewers *what* to think about and *how* to think it' – Bauman, 2002, p. 66). Bauman regards *Big Brother* as a laboratory for a very public experiment in eliminating spontaneity and turning human personality into a thing, the sort of experiment that Hannah Arendt identified as being undertaken in the concentration camps of totalitarian regimes (2002, p. 67).

The possibility that *Big Brother* is neither reducible to a simple reflection of reality nor an injunction of how to live, and that its viewers (other than himself) might be anything other than cultural dopes, does not seem to occur to Bauman. As John Ellis (2009) tells us, reality television reinvented participatory television by combining elements of a game show and documentary. It is a game show that belongs in the tradition of Saturday night televised entertainment in which ordinary people take part in games devised by television companies, usually with a host such as Michael Miles, Hughie Green or Bruce Forsyth (Ellis provides an evocative account of Hughie Green's style of tele-visual sincerely felt emotion – 2009, pp. 105–7). There is no genial host on *Big Brother* because the dramatic conceit of the show is that it is us, the viewers, who decide who stays and who goes; we are the big brother. But the familiar format continues a ritual function in which viewers are brought together to watch what happens and to participate in the voting process. This is not mediatised ritual in Simon Cottle's sense (2006), but it is a ritual event created by the media, which has a repeated and habitual form with a distinctive moral purpose. As Ellis explains, quoting Annette Hill (2004), the reality show itself does not provide moral evaluation, but it does display a range of moral issues, including privacy, rights to fair treatment, good and bad moral conduct, and taste and decency (2009, p. 111). I have argued that all television shows human social action and its consequences in a way that allows the audience to reflect and remark on it. The appeal of reality shows – and *Big Brother* is a key example – is that they show people interacting without a script so that what is shaping what they say and do is the combination of the setting, the unfolding situation and the personal-ities involved. This gives viewers an opportunity to watch how person-ality traits are manifest, how stratagems or interactive tactics work, how ways of putting things in speech and the use of nonverbal communi-cation lead to responses and how rapport is established or undermined. The viewing pleasure is in watching these facets of interaction, the same pleasure as 'people watching' which is part of all public social settings, with the added advantage of being able to watch closely and hear what is said. There is the further strange advantage of not being able to get 'caught' overhearing a conversation because the participants have given their prior consent.

Viewers are in a position – just as they are when they watch social interactions in face-to-face social settings – to make moral judgements about who is acting appropriately or well, who is being fair or kind and who is being selfish or insensitive. The moral import of what is seen becomes the theme for a series of other media content – tele-vision shows, newspaper features, Internet chat – and in those face-to-face interactions people have at work, at home and at leisure. As Ellis put it, 'Reality TV enables social talk about moral values and about how to understand human behavior' (2009, p. 111). Ideally, this reflective judgement on the morality of the participants would be linked to the judgements made in telephone voting, but of course there is often more pleasure in watching excessive or bad behaviour than watching those who get on easily with each other. Those who are too easy-going are unlikely to try to become a participant, are not likely to be selected if they did, or are voted off early on in the show because they are deemed dull and 'uninteresting'; their actions do not provoke controversy or censure. Successful players are often those with strong, identifiable personality traits whose interactional performance was sustainable and worked in the rather strange social context of the *Big Brother* house. The need for contestants to be quirky and distinctive, to want the sort of brief public celebrity that comes with being on television for a sustained amount of time, has limited the show as it developed from season to season. The show's producers and those who applied to be contestants became more experienced in recognising what made the highpoints spontaneous and memorable, but when they tried to recreate or reproduce them they undermined their humanity and spontaneity.

Witnessing

The possibility that television presents or represents 'reality' is problematic in a way that is compounded by the obvious contrivance of so-called 'reality TV'. A different approach to the capacity of television to mediate the world we live in is captured in the idea of the medium as a 'witness'. John Ellis introduced the term as he tried to grasp the importance of the photographically precise quality of television's visual evidence that provided '...quasi-physical documentation of specific moments in specific places' that 'brought us face to face with the great events, the banal happenings, the horrors and the incidental cruelties of our times' (2002, p. 9). For Ellis, it is the relative inability of photographically exact images to distinguish between subject and setting that give the camera its evidential force, and the implicit claim to liveness of television imagery increases its potency as a witness (see Chapter 5). This means that irrelevant and unimportant details are included, such as the slogan on the clothes of a disaster victim, the décor of a domestic setting and ambient noise that are all captured whether or not they are relevant to the story being told. In live, actuality television, the programme makers have very limited control over these elements in contrast to recorded, dramatic settings where sets are 'dressed', lighting is controlled and unintended ambient sound is excluded. Derrida discusses the question of televisual testimony and in particular the footage captured by a witness of the beating of Rodney King that, because it was replayed on television many times, had a very powerful effect. He brings together the issues of witness and the truthfulness of the medium when he says:

> No one could look the other way, away from what had, as it were, been put right before his eyes, and even forced into his consciousness or onto his conscience, apparently without intervention, without mediator. (Derrida and Stiegler, 2002, p. 92)

What impresses Derrida here is the power of the camera, an impersonal technical device, to give testimony as a witness (although legally it was the cameraman who was the witness) to what he describes as a banal event, compared to the much worse things that happen and never get onto a screen.

In a metaphor that plays on the way the city life of modernity brought people into much closer involuntary social proximity, Zygmunt Bauman has called the capacity of television to provide viewers with the ability to witness the lives of people whom they do not know, the 'telecity':

> Strangers may now be gazed at openly, without fear – much as the lions in the zoo; all the chills and creeps of the roaring beast without the fangs ever coming near the skin. Strangers may be watched robbing, maiming, shooting and garrotting each other (something one would expect strangers, being strangers, to do) in the endless replay of TV crime and police dramas. Or they can be gleefully gazed at in the full flight of their animal passions. (Bauman, 1995, p. 178)

In the telecity, we encounter strangers whom we can control as objects (switch channels, ignore them, vote for them, turn the set off), but they are not subjects in our lives and we have no responsibility for them, no moral relationship with them. We have no obligation to care for the characters we see on the small screen, whether they are real or not, because they are taken care of by the television company. But Ellis's idea of witness is summed up in the idea that 'You cannot say you didn't know', which he attributes to the newsreader Trevor McDonald, in which the media audience become accomplices, obliged to do something, if it is only to react. During *Big Brother 5*, a fight broke out following Michelle and Emma's return to the main house, and alarmed members of the public who were witnessing the violence on screen called the police, who then investigated the incident (Plunkett, 2004; Holmes, 2008). Complaints from 55 viewers led to the show being referred to the regulator Ofcom, who later ruled that the coverage of the drunken brawl breached its programme code (Deans, 2004). The call for action is an element of media witnessing that follows from you-cannot-say-you-didn't-know in which the veracity and immediacy of televisual images, evidenced by the liveness of photographic audiovisual imagery, engages at least some of the audience into feeling responsible and that they should act.

Witnessing is a complex process, as John Durham-Peters (2011) points out, in which there is a mixture of the courtroom evidence of the 'eyewitness' who gives testimony as to what they saw but not their interpretation or emotional response, and the religious witness who attests to their inner experience rather than what they perceived. What is significant about witnessing in general is not the perceiving or the inner experience so much as the recounting of it, the testimony that is offered to others in a declarative mode as evidence. We associate the notion of witness with the agency of human beings and their capacity not only to perceive but also to recall the evidence of their senses and discursively summarise them – those whose capacities are lacking, such as through youth or extreme old age, are deemed less reliable.

Television does produce people who are witnesses – reporters, bystanders, those who were involved, professionals and experts – especially in actuality programmes and documentaries. But the apparatus of television itself functions as a witness through its capacity to reproduce what its equipment 'sees' and 'hears', and since it does not interpret what is captured, it is not partial or selective but abundant to a superfluous degree. There are, of course, human agents involved behind the camera and microphone whose importance, though often overlooked, is in operating the equipment but above all in being there and choosing to capture *these* events. Their agency gives the medium itself a role as a witness and throws into relief the obligation of the viewers as witnesses, albeit at a second degree. The viewer also sees and hears, they are quasi-present at the event that is channelled through the medium. Here is where the 'reality', the mixture of veracity and verisimilitude of the audiovisual image, reasserts itself so that the viewers cannot deny or ignore what they see and hear.

Durham-Peters argues that there are different degrees of responsibility that go with witnessing according to presence and liveness because there is a 'veracity gap' that is widened by distance. Those who are present in space and time, like those who passed the injured Samaritan, are most complicit in what is going on; those who are not present in space but linked through a live transmission simply feel an obligation, while those whose experience is distanced through space and time by a recording are even less involved (2011, p. 38). This is disputed by Paul Frosh (2011), who points to the importance of the intermixing of the human witness – often a journalist but also the equipment crew – with the audiovisual imagery that blends the experience with its telling. The discursive authority of the journalist anchors the events in space and time, giving them a narrative context as well as testifying to the veracity of the images. For Frosh, this is more important than the distance in space and time of the mediated witness and is even 'morally enabling' because it maintains an equivalent distance amongst the strangers who are viewers. The idea of a 'veracity gap' suggests a limit to the truthfulness of what we can know but Frosh reminds us that witnessing is not only about representing the truth but also about the cultural achievement of communication, about giving testimony, which is not only a technical but also a rhetorical matter. Some witnesses are, as any lawyer or evangelist knows, better than others, and the makers of television programmes are sensitive to the link between showing and telling. Paul Frosh links Goffman's concept of civil inattention to the globally-mediated cosmopolitan culture of late modernity to point to this fundamental moral role for media witnessing:

> As a moral force, then, media witnessing – like civil inattention – is a routine and institutionalized social procedure for moralizing strangers by placing them within the framework of those whom we recognize as equally human. It extends to them through its very impersonality and generality the fundamental non-hostility and equivalence that underpins both modern cosmopolitanism and the discourse of universal human rights. (Frosh, 2011, p. 69)

A number of different television programme formats borrow the documentary device of witnessing the actual setting to connect an audience with a culture very different from their own. The history program in which an authoritative presenter strides through the original settings of significant events, telling us about what happened, interspersed with dramatic reconstructions, is a recurrent form (for example, *A History of Britain*, BBC One, 2002, *Monarchy* Channel 4, 2004–5). Others are the travel show where a celebrity moves through a different country encountering and commenting on its culture, both material and performed (see Chapter 3, p. 57 for examples). All these programmes use audiovisual imagery to witness a 'different' culture and an experienced and suitably qualified presenter as a witness to bridge the veracity gap (for example, Simon Schama, David Starkey, Peter Owen Jones, Bruce

Parry). These professional witnesses develop a televisual persona that carries the blend of sincerity, authority and articulacy that will attract the audience's trust in the truth of their testimony. They attest to the taste of the food we have seen prepared and cooked, to lives lived long ago in the buildings, paintings or other artefacts we have been shown and to the authenticity of local characters and scenes that unfold on the screen.

A departure from this mode of professional televisual witnessing was *Blood Sweat and T-shirts* (BBC 3, 2008), in which six young people who had no track record of television work or academic status to give them authority as witnesses, were sent to India with TV cameras following them, to learn about the way that clothes are made. Their qualification was their regular consumption of high street fashion, and their role was to 'be themselves' as in a reality format. They embarked on the trip with confidence and enthusiasm, willing to learn and keen to join in, and their emotional journey of coming to terms with the cultural difference they experienced made them potent witnesses of the hardship of those living and working in the garment industry in India. In the first of four episodes, the density of the population, the smells, the cramped accommodation and the work ethic shocked them. They were taken aback by the expectation that they would have to learn their work tasks quickly and under a strict regime of discipline. One young man, Mark, did not like being told off for turning around and talking, and later he got very upset when the floor manager touched him while trying to explain how he should use the foot pedals to control the sewing machine. There were tears and outbursts from two of the girls over the pressures at work, and the group were upset when they realised that they had unintentionally offended their hosts.

As witnesses of the different mores of home and work life, the six young people provided a set of responses that were emotional and indicated their limited experience of the world. It was as if they have not watched enough television about life and work in India; for those who had, the situations shown on the screen were not so surprising. But of course, they were actually there, living within the difference, experiencing it through all their senses, 24 hours a day. And over the series, they became culturally acclimatised and found a way of being within this set of social contexts that was very different from those they were used to. As they reflected on the lives of the people they met, they accepted that they could not judge the moral order of this different society by their own lives in the UK. Tara, who aspired to work as a fashion designer like her mother, said: 'The dilemma that I've got is that I really can't condone the conditions of some of the factories that we've visited but I can't really condemn them because they are providing people with jobs and with roofs over their heads'.[8] The moral issue that rapidly emerged as a focus for both the programme makers and the young people, was the exploitation of child labour; they found evidence of child labour in some of the factories they visited and in the last programme accompanied an enforcement officer as he searched for children in factories and visited a school for rescued child labourers. One of the six, Stacey Dooley, went on to make other programmes that focused on the exploitation of children in different types of work in different countries. She had established her capacity to be a televisual witness during *Blood, Sweat and T-shirts* through her empathy with the people she met, her persistence and her articulacy. For television to be an effective witness, it needs to have a visible human agent whom the audience can trust and identify with. The capacity to express emotion is important, but if it is overwhelming, it draws the attention to the presenter and away from what they are witnessing – as happened in the early episodes of *Blood Sweat and T-shirts*.

Conclusions

The small screen can *be* moral by how it shows us what it does – this is the responsibility of those who make television. As well as simply showing us moral behaviour that we viewers can judge as impartial spectators, those who produce and broadcast television can have a moral effect through what they choose to show. I have argued that there are a number of different modes in which the medium of television can be moral.

Firstly, those who make programmes have professional ethical standards specifying that they should tell the truth and not exploit people. The audiovisual impression created on the small screen should be the truth in the sense that it is what any reasonable person would recognise as the truth. This is not an epistemological criterion but a practical one, such as we would expect in ordinary life; where what appears on the screen is fictional, its claims to represent the real as it was, is, or might be should be clear to the viewer. We can expect that television does no harm, that is, it should not cause suffering to anyone it portrays. There is a presumption that the truth cannot be harmful – though a journalistic ethic may be needed to judge whether harm to individuals is, or is not, in the public interest. The ethical standard of truthfulness refers back to the morality of the culture from which the televisual imagery arises, but there is also a quasi-judicial aspect to maintaining the standard of truthfulness that in the UK is regulated by Ofcom.

Secondly, television may become caught up in a moral panic, such as the concern about swearing that was linked to the showing of sex and violence. Television makers have to make judgements about the interests and tastes of viewers in general while avoiding getting caught up in the media amplified concerns of a few moral entrepreneurs. At the same time, they need to be aware of changing mores about what sort of behaviour can be shown when, and in what sort of context. A third way in which the small screen can be moral is through showing the distant suffering of other people, which, were those people close socially and geographically, would demand some sort of action on the part of the viewer. The audiovisual depiction of suffering is particularly powerful in evoking an emotional response of care and concern in those confronted with it; they may choose to sacrifice time or money to help, they may denounce those who they see as complicit in causing the suffering, but they may become distracted by the aesthetics of being moved by what they see.

Showing what is true and showing distant suffering gain their moral power because of the capacity of the small screen to show what is real. The fourth mode of televisual morality is in representing the real, making what appears on the screen 'realistic' as well as factually true. The perceptual realism of television can sometimes be the basis for accepting the truth of what is shown, but audiences become sophisticated in distinguishing what is made to appear real, from what is actually real. Nonetheless, the reality of what is shown may be compromised by the realistic fakery of earlier imagery, as the WTC disaster was by decades of disaster movies and television shows whose narratives follow a depiction of disaster. A distinction between the real and the Real emphasises the importance of shared ideas about how the world must be, beyond direct sensual experience. Those shared ideas include a sense of what is appropriate, how things should be and the taken-for-granted moral order of viewers' culture.

Fifthly, one way in which the real is made real is through the capacity of television to show and to be a witness. Truthfulness here is not simply a perceptual capacity but a property of testimony, the recounting or retelling of what is perceived. The witness role of television, especially the actuality content of news and documentary programmes but also 'reality' television, can be routine and familiar, what Ellis (2011) calls 'mundane witnessing'. Even when the emotions of viewers are not provoked by what they witness on the small screen, even when they remain indifferent to what they are being shown and what is being related, viewers are participating in a moral culture that accepts the equivalence of others.

8
Television and the Imaginary

From deliberation to imagination

To act according to rules is to have disciplined behaviour and might be seen as ethical, but to be moral requires having an imagination through which to consider the possible consequences and ramifications of an action and so distinguish between what would be right and what would be wrong. It takes imagination to reach a judgement that can inform one's own actions or react or respond to other people's actions. This does not necessarily mean that moral reasoning requires a rational, conscious, cognitive act of weighing up pros and cons, calculating causes and effects. In the imagination, a person may *see* what the outcome will be and have the sense of simply knowing what is right or wrong. This capacity of the imagination often becomes habitual or routine and is the result of prior experience, of having learnt through seeing similar social situations followed through. Some situations may require a more conscious or semi-conscious process of deliberation and reasoning, especially when decisions have to be made collectively as, for example, in making a decision about the ethical consequences of how to treat an ill patient. For collective decisions, a discursive sharing of the process and the judgement will be necessary, but for the individual actor, moral decisions will usually be silent and intuitive. There is an emotional component to the individual process that makes it more fluid than the mechanical calculation of consequences, so that the person *feels* what would be the right thing to do.

What is imagined need not lack reason or deliberation, and it may be founded on empirical experience or prior rational judgement, but the imagination can never be reduced to reason. The words 'imagine' and 'intuition' are both derived from Latin roots that refer to the human sensuous capacity for seeing, gazing, looking on, representing and picturing mentally. When we talk of 'imagination', we are often referring to seeing pictorially in the mind, although the image we hold may have other qualities such as good or bad feelings that are evoked or associated. The picture does not need to be complete as a representation to generate such emotional responses, although often it carries traces that mean we can explain to others, or ourselves, the reasons that lead to the emotions that are stimulated by our imaginings. Judgements are not only shared through the exchange of reasons and through the conscious, outward process of discussion, they are also shared through images that are held in common. One of the features of morality is that it can often seem obvious what the right thing to do is or would be... and yet another person would do things differently.

The idea of the 'imagination' seems to suggest something internal, cognitive and somewhat private because I can imagine something that no one else knows anything about. But even this private projection in the mind of something that is apparently independent of our senses always derives from what we have learnt from our engagement with the society around us. This social quality of the imagination is, I will argue, what constitutes an 'imaginary' – dynamic images held in the mind that are broadly shared with other people. These are not like a series of photographs or paintings because they have consequences and outcomes that are part of the image. What I can conjure up in my mind as an image is derived partly from my experience and partly from the cultural context that has associated values and meanings with that experience. These values and meanings may seem to be mine, but in fact they are shared, if not by everyone, then by many others. And it is through these valuations and associations that morality comes alive in imagination, leading to judgements about actions that have been performed well and also about those that simply appear possible.

Let me explore the idea of imagination with something that is familiar and apparently lacking in moral consequences: imagine a sandy beach with waves from a gently lapping azure sea and a warm sun shining from a blue sky with some white fluffy clouds floating near the horizon. Two qualities of this fantasy are that nothing else needs to happen, and the pleasant environment can endure for as long as one wishes. This very corny fantasy represents experiences I have enjoyed a number of times in my life and is informed by the sensations I recall from them – I hope that you have also enjoyed similar experiences. The scene is one that I might draw upon to distract me from something that I am not enjoying but have little control over, such as being in a dentist's chair.

However much the image rests on my own sensuous experience, it cannot help but be informed by vicarious experience of similar situations – similar beaches that other people have described, the photographs they have shown me, the postcards they have sent me. It is indeed a stereotypical representation of the idyll of relaxation, of being without concerns or pressures. It is familiar to many people – even those who do not particularly enjoy beach holidays – as an image that evokes pleasant sensations. The imagination is mine, and the image is within me, but the values and meanings it has of relaxation, of not being at work, of free time, of comfort and ease are culturally shared. And these shared associations enable the image to be a stereotype, one that is used as a base in many films and television programmes as well as the images in holiday brochures and advertisements. It is the shared nature of the image and its meanings that enable it to be part of what we might call 'an imaginary' – a collection of dynamic images that have meanings, associations and values that are broadly shared, or at least understood, by the members of a society. Something that is important about this way of approaching what I am calling an 'imaginary' is that it is primarily made up of pictorial representations and feelings. It is not composed primarily of discourse, of sequential, abstract forms that are combined to carry meaning. It may be that an imaginary is stimulated by discourse or represented or referred to through language. But it is important that it is not reducible to language; it is always already extant as a dynamic image.

There are, in fact, moral implications of my image of being on a warm, sunny beach (such as the associated feelings of lack of responsibility or care for others or for the future) but not ones that are of great importance – they are certainly not ones that affect the image being useful to help me relax in the dentist's chair. In this chapter, I want to explore further how individuals' imaginations are shared through collective imaginaries and how this is connected to the maintenance of a moral order that is characteristic of society. The idea of a 'moral imaginary' as a shared repository of moral attitudes and responses derives from a number of sources that are worth expanding on a little before arguing the case for television's role. The theme I want to develop is that television contributes to a 'moral imaginary', by which I mean a pool of ideas and images about how people do behave, how they should behave and what the consequences of good and bad behaviour are. Ideas and images do not exist as concrete and material forms that could endure as a library does – both the building and the books it contains – but they do exist in the minds of people who share them through communication. It is through the continual sharing of instances of appropriate and inappropriate ways of acting that a moral imaginary is sustained within the minds, memories and habits of the people of a society. The moral imaginary is at once a shared repository of these images and ideas, while at the same time it is a cauldron in which they are reviewed, modified, developed and reordered.

The spectacular

What is fundamental to the concept of a 'moral imaginary' is the nature of morality as a set of interconnected dynamic images that are not reducible to a narrative form and whose dynamism plays in the minds of the members of society. The reason why television is such a powerful device for fuelling the moral imaginary is that it *shows* morality as moving, unfolding images. Narrative may be important, but the play of moving images is even more effective in exposing a population to a wide range of moral ideas. Thinking within the moral imaginary is not only logical, rational or calculative – it is also affective, visceral and, above all, visual. I argued in Chapter 5 that it is important not treat television as a medium that is 'read' because the encoding of ideas and experiences in written text that we read involves a process of decoding which is substantially absent from continuous moving images relayed or recorded through a camera that is in front of real events. There is often text displayed on television screens, which has to be read, and there are conventional forms – graphic images for example – that do operate as a code that also has to be read. But here I want to introduce the idea of the television viewer not as a reader but as a *spectator*.

Television viewing is generally rather passive, and viewers engage with what appears on the screen much as spectators in other contexts – at sports and other games, theatrical events, festivals, circuses, shows, regattas and even watching the night sky. A spectacle is often an event that is either impressive as a natural form – the waterfall, the storm, the range of hills or the colourful sunset – or a human creation that is intended as a display; the air show, the *son et lumière*, the firework display or the pantomime. The spectator is an observer, someone who watches as a bystander rather than a participant, an onlooker who is interested and who may become a witness to whatever event or spectacle it is they see before them. The content of television may on occasion be spectacular in this way, such as the live televising of the opening of the Olympic games or the celebrations that surrounded the millennium in different capital cities (see Lury, 2005, pp. 127–46). However, I want to suggest that even ordinary, routine television puts the viewer in the role of a spectator gazing from a static position as events unfold, but the passivity is simply of the body, whereas the mind, the emotions and the imagination can be stimulated and engaged.

Guy Debord's (1983) provocative account of the *Society of the Spectacle* re-thinks Marx's sociology for a world in which representation has taken over a fundamental role. As he famously puts it, 'The spectacle is not a collection of images; it is a social relation between people that is mediated by images' (Debord, 1983, 1 § 4). He argues that representation, whether in the form of information, advertisements, propaganda or entertainment, takes over the social relations outside those of production, not as an alternative mode of existence but as the form of reality in modernity. Debord claims that 'the spectacle is an *affirmation* of appearance and an identification of all human social life with appearance' and that critique exposes it as the 'negation of life' (1983, 2 § 10). Appearance has become the mode of being as well as the mode of power in late modernity, he says, and the spectacle is a form of the accumulation of capital and a means by which people are alienated from their conditions of existence. For Debord, the transformation of modern culture into a society of the spectacle leads to the banality of consumer culture in which the proletariat are politically passive as spectators of history, unable to recognise and act in their own interests. The imagery, he argues, transforms culture into a commodity and one that is trapped in the past and resistant to change.

Debord's declarations do not identify television as a distinctive medium in the constitution of the late modern spectacle. Debord merely treats it as another consumer object distracting people from their real situation. But while television does not lend itself to the spectacular because of its small size and mundane presence, it is a prime vehicle for the society of the spectacle in which images fill the communicative space between people. What I wish to argue is that through the 20th century modernity became the society of the spectacle through an increasing emphasis on visual communication of which the television is a key part. But I think Debord is wrong in presuming that it necessarily leads to banality or alienation because there is no overarching process that denies the agency of the spectator who can always respond to images in a variety of ways. The audiovisual moving image is a mode of communication that includes and shares without any compulsion to accept any particular images or attribute particular significance to them. The interests of capital may control the channels that feed our small screens, but they do not have a coherent interest in morality beyond simply maintaining a moral order. The reason I think that Debord is wrong is that his concept of the image is too simple and overly deterministic and lacks an understanding of the human imaginary – I will develop these concepts through the work of Jean-Paul Sartre (2004), Jacques Lacan (1977; 1979) and Cornelius Castoriadis (1987). Then I will explore the agency of the spectator to argue that it can create what Adam Smith (1976) called the 'impartial spectator', and there is potential for what, more recently, Jacques Rancière (2009) has called the 'emancipated spectator'.

The imaginary and the analogon

The rendering of the social order as a series of appearances on television has contributed to a shared collection of dynamic images that I am calling an imaginary. Because many people watch the same programme at the same time, they are able to refer to it in casual conversation, showing that they were 'there', as the tragic situation of miners trapped on the other side of the world unfolded, for example, or were impressed and moved by the moment of a spectacular goal scored by a member of their national team. What is on the television is not spectacular itself, but the presence and visuality of the medium can evoke spectacle and the illusion of co-presence like no other. Its images become part of the way that we see the world, recognising people and places because we have already seen them…on television. The word 'imaginary' works easily in ordinary English as an adjective; it describes that which does not really exist, or at least things that have no material form. It works particularly well to describe things that have never had form but can be imagined as if they did. The use of 'imaginary' as a noun is less usual in English, but it has emerged through usage in French by authors such as Sartre, Lacan and Castoriadis.

Jean-Paul Sartre's (2004 (1940]) exploration of the imaginary is from the perspective of phenomenological psychology, in which the intentional human subject engages with images. Sartre sees the imaginary as the holding of images in consciousness, but he recognises that an image is a relation to the object: '…it is a certain way in which the object appears to consciousness, or, if one prefers, a certain way in which consciousness presents to itself an object' (2004, p. 7). He distinguishes between the textual sign that is read and the image that is perceived. The individual letters and words of a text are meaningless in themselves and rely on convention and habit, but 'Between the physical image and its object there is a very different relation: they *resemble* each other' (Sartre, 2004, p. 21). In the image consciousness, we 'comprehend an *object* as an 'analogon' for another object' (Sartre, 2004, p. 52) that stands in for it rather than merely representing it – it is the phenomenological idea of 'intention' that orients the imagination towards the original object or person and enables it to be imagined. Writing of a portrait, he says that the object is a *'quasi-person,* with *a quasi-face'* that may on occasion even be taken, for a moment, as the real person and at least carry the expression of the real person (Sartre, 2004, p. 22). Sartre discusses different forms of imagery – portraits, theatrical mimicry, line drawings, hypnagogic images – and argues that behind all these forms is an intentionality that aims at what is being represented and not at the material object that is thereby created. No image is ever a perfect analogue of the thing it resembles, and there must also be an 'imaging consciousness' which perceives the material object and interprets it according to existing knowledge to create a mental image. The creation of this 'equivalence' depends on *'symbolic movement,* which, by its very nature as movement, is on the side of intuition and, by its signification, is on the side of pure thought' (Sartre, 2004, p. 51).

The imaginary is that realm of consciousness populated by images that are *analogons* of the contents of perception. The analogon is closer to the process of perception, a more visceral engagement, than that of encoded knowledge such as that in the form of language or an algorithm. The 'psychic form' of an image – say, of my daughter – may be a compound of real images, none of which matches a particular photograph, and which is difficult for me to inspect in the way I would a photograph, by scanning across it. The psychic image is nonetheless related to how I perceive her when she is present to my senses and would, for example, be involved in my recognising her in a crowd. The details of my mental image of her enable me to interrogate it so that I can see different aspects at once – the colour of her eyes, the shape of her ear – and even see her dynamically in movement. The imaginary may include the sound, the smell or the feel of something or someone, but it also includes knowledge and emotional orientation: 'So knowledge aims at and affectivity reveals the object with a certain coefficient of generality' (Sartre, 2004, p. 90). Sartre distinguishes perceptual observation, when the object is real and present in relation to the body and its senses, from when it is 'quasi-sensible' when the imagination creates an image of an object or a person who is absent but was perceived in the past (2004, p. 125).

The television provides us with images of people who appear on it so often that we can call them to mind in the same sort of way as we do the faces of friends and family members. The capacity of the mind to synthesise past observations into current images means that we can conjoin various events that were not precisely the same. The images shown on television provide a 'quasi-sensible' stock of images that the intentional consciousness can work with because they are moving and taken from a number of different angles. This can lead to confusions such as viewers watching a repeat and being unsure whether they have seen it before, or when an actor from one show appears in another, and the two characters overlap in a viewers' mind.

'Irreal' objects draw on our previous sensual experience but have never existed; they are created through the spontaneous intentional operations of the imagination. The centaur, the Greek mythological character with the top half of a man conjoined with the body of a horse, is among the irreal objects that Sartre uses as an example of what can be held as an image in the mind that doesn't age or change (2004, p. 132). The centaur is an image with particular characteristics even though there is no 'real' object of which it is a depiction. It can be drawn, used as a symbol and even 'brought to life' through the animation techniques of computer graphics, as in the films of C. S. Lewis's *Chronicles of Narnia* stories. In the imagination, irreal objects and people can be manipulated and perceived as if they were real – Sartre uses the example of committing a murder – but only by an irreal self (2004, p. 125). Whatever happens to these things arises through the intentionality that drives consciousness; the blood flowing from a murder victim, the way they fall, can all be brought to mind by intention even if the event were never witnessed. But this does not mean that the irreal image has no effect on the person who can feel love, hatred, admiration, disgust and other emotions towards it through 'a kind of continuous autocreation, a kind of restless tension' that inflates the emotional response to what is imagined (Sartre, 2004, p. 140).

Television provides readymade irreal objects that we perceive, can link values to and experience emotions about and then retain as images, sensations and concepts in the mind. The romantic drama moves us with images that are associated with feelings of love and affection, hospital shows – whether documentary and actual or dramatic and fictional – often provide us with images that are linked to fear and apprehension. As a televisual experience, we enjoy the play of emotions in what we see, often stimulated by sentiments expressed in the script. While we are enjoying the experience of the emotions evoked by these characters and their actions, we are also learning to refine just what emotions we might feel and how we might feel them. As we make sense of what we see on the screen using the same resources as we use to make sense of reality, we might not notice the symbolic enhancement of irreal objects that is achieved by the craft of televisuality. This may be the music that accompanies actuality programming almost as often as it does drama, but it also includes the camera angle and its proximity to people and events, which are unavoidably part of the representation of reality. For example, when a person discussing the illness of their child in a medical documentary is shown in a close-up, we can scrutinize the expression on their face that communicates very real emotion without the constraining embarrassment we would feel if we were physically co-present. The feelings of viewers are stimulated through exposure to these screened images, which then provide the imaginary with a stock of irreal images that include the associated emotions.

Sartre points out that the imaginary life involves a return to images and feelings, to reliving a love life in the imagination or revisiting the horrors of a past experience. He reminds us that there are 'imaginary feelings' to do with what is imagined, which should be distinguished from genuine feelings which are a response to a real situation, co-present in time and space with the inexhaustible complexity and richness of real people. Imaginary feelings are relatively simple and banal, degraded because they conform to fears and desires (Sartre, 2004, p. 145). Viewers seldom have difficulty in distinguishing the imaginary feelings they have for characters and situations they perceive on the small screen from genuine feelings, but they will seek out viewing experiences that evoke certain feelings. The warmth and familiarity of characters who act out imaginary situations in a way that endlessly confirms their warmth and familiarity is the staple format for long-running series in which all that is left to enjoy is the wit of the script writers in weaving in jokes and creating 'new' situations through which to reproduce the same responses.

The classic of this form is the American situation comedy *Friends* (NBC/Warner) that originally ran for ten seasons between 1994 and 2004. The humorous interplay between characters in the 236 episodes provides for ups and downs and twists and turns in which sadness and hurt are always countered with amusement and humour. Despite the whole series being available on DVD for over fifteen years with many shows and extracts available on the internet, the show has been endlessly repeated on broadcast television (thirteen different episodes from four different seasons are playing on three different channels in the UK in the week in which I am writing this – some of these broadcasts are further repeated within the same week on timeshift slots). The show creates a series of irreal objects in the hugging and mugging and the flow of double-entendres, repartee, non-sequiturs, pratfalls and farcical entrances that give the illusion of a spontaneous and lively imaginary world that would be warm, safe and amusing to inhabit.

The imaginary may contain images derived from literature and story telling but those media demand that the individual creates them for themselves. In contrast, the moving audiovisual media provide full, dynamic images, often from many perspectives with much extra and contextual information, such as realistic movement and ambient sound. Their fullness means that they are ready, available and accessible as irreal entities that constitute a shared imaginary for many members of industrialised societies. Sartre discusses hallucinations, hypnagogic and dream states that involve the imagery coming to mind in ways that are similar to the operation of the imagination. These less-than-conscious states of mind may also be stimulated by, or draw on, the contents of a shared imaginary that television has contributed to. The rational mind that directs thought and imagination has no purchase in these states, and yet they are opportunities for the re-experiencing of feelings and attitudes, the refreshing and enlivening of quasi-sensible and irreal objects. As Sartre says, the dream world is that of a 'story', lived as a fiction, 'the odyssey of consciousness dedicated by itself and in spite of itself to building on an irreal dream world' (2004, p. 175). The odyssey often confuses the temporality of the objects in the imaginary in much the same way as does the delight of television in showing repeats and replays, even of topical comedy shows that are now out of sync with the news items they make jokes out of.

Watching television has much in common with the dream-state or hallucination in that viewers have to 'suspend reality', the consciously experienced paramount reality of the everyday world, to give themselves up to the story and the lives of the people represented, whether fictional or actual. Even the bodily responses accompanying these states that Sartre describes – flickering eyes, twitching hands, raised heartbeat, arousal – are reminiscent of those that come with watching screens. However, the difference from a dream or a hallucination is that we can reflect on the moving images on the screen as we see them, discussing what happens with the person watching with us, or engaging in a continuous dialogue with ourselves about how we feel about what we are seeing, how beautiful, exciting, dull or depressing it is making us. We can even recognise the 'moves' of televisual devices and techniques (for example, the camera zoom, the disembodied commentary, the amplified sound effect) that stimulate our responses while still feeling those responses. Sartre discussed how those images that are seen by many people, contribute to an imaginary, but because knowledge and affect also shape the irreal object, it will be different for different people.[1]

The specular image and the imago

To help understand the constitution of human consciousness, the psychoanalyst Jacques Lacan (1977; 1979) coined a tripartite distinction amongst three realms of experience through which the individual and their identity is constructed: the imaginary, the symbolic and the real. His English translator helpfully describes the Lacanian imaginary as 'the world, the register, the dimension of images, conscious or unconscious, perceived or imagined' (Alan Sheridan in Lacan, 1977, p. ix). The realm of the imaginary takes shape in Lacan's (1977, pp. 1–7) writing about the 'mirror stage', the moment sometime between six and eighteen months after being born that marks the entry of a child into society. The child recognises the person reflected in the specular image in a mirror as himself or herself gazing back, an independent human being with an exteriority separate and alongside other humans. This image of the specular self is distinguished from the *imago*, a psychoanalytic concept of the idealized image of a loved and important person, which is carried within someone as part of his or her emerging individual identity.

The *imago* of a parent is a familiar and recurring image of a whole separate person with whom there is a continuous emotional connection, distinct from oneself and other beings who are perceived as also human and sharing similar motor, cognitive and perceptual capacities. For Lacan, the identification with the *imago* of idealized parents and the reflected self – including the people and surrounding objects – each works as a *Gestalt*, as a whole form that has meaning in itself (Lacan, 1977a, p. 2). These images contribute to the coming-into-being of the subject, giving it agency that is irreducible and prior to the language-based identification with the other. What is interesting about Lacan's focus on the young child and its orientation to the world around it is that it is pre-linguistic, operating with emotional and intellectual depth based on imagery. The *imago* is an idealised three-dimensional image of a whole being that moves and endures through time, as against a static, two-dimensional icon. The *imago* is recognisable from all angles, regardless of how it is dressed or the expression on its face. Images captured through the gaze provide the basis for recognising people as people, and significant people in particular, and the end of the mirror stage signals the preparedness of the individual to begin to move from a 'specular *I*', fascinated by its own image into a 'social *I*' (Lacan, 1977, p. 5).

Lacan's concept of the 'imaginary' is a resource for the individual subject as it engages with the world, but it also provides a basis for understanding the priority of the image over the symbolic. It is a theme that he returns to as he discusses the role of the gaze, mimicry, the mask and the screen in how seeing also involves being seen (Lacan, 1977b). The gaze, the resting of the eye on a form that is noticed as a whole, as a *gestalt*, is a feature of Lacanian psychoanalysis that has led to his being taken up by film theorists. Recognising the human form or *imago* as special is an early part of learning what is important in the world. If Lacan's imaginary is the relation between the ego and recognised images, the symbolic is the relation between the subject and the signifiers of speech and language, while the 'real' is the realm that images represent and which symbols invoke, both always more or less inadequately. The imaginary for Lacan includes the illusions and deceptions that the psyche creates for itself, but there is every reason to believe that it is also susceptible to the illusions created by others – such as those displayed on a flickering small screen. Lacan's psychoanalytic approach gives a sense of the power of images in the life and identity of the individual and in particular their embedding deep within the person.

The psychoanalytical concept of the *imago* refers to significant others, usually parents, but it is not difficult to see how a similar process operates to idealise the dynamic image of the celebrity. Just as paintings and photographs of loved ones become part of an *imago*, so repeated dynamic images of celebrities enter the shared imaginary. The celebrity is someone who is easily recognised, whatever they are wearing or saying, who is worthy of attention and even emotional attachment. We have the illusion that we *know* the person whom we see frequently playing the role of singing star, grumpy detective or serial seducer. Their persona is perceived through the audiovisual sensations we receive so that knowledge and affect become tied to the intentional image we retain. When we see in our local supermarket the actor who plays the grumpy detective, there is a moment of recognition before we can distinguish the detective persona, whom we know at the level of the imaginary, from the person, the actor buying soap powder, whom of course we don't know at all. Celebrities are usually celebrated for particular qualities – the sportsman for winning a tournament, the actor for an acclaimed performance – but once they have become familiar, public interest extends to their opinions and experiences, their domestic and private lives.

Before the society of the spectacle, it was only royalty, music hall performers and perhaps civic dignitaries who appeared sufficiently often in public to be idealised as celebrities, but the visual media have filled the shared imaginary with many more quasi-imagos of 'B' list celebrities. The celebrities who appear on television quiz shows and reality games such as *I'm a Celebrity Get Me out of Here* (Granada 2002 licensed to France, Hungary, India, the Netherlands, Sweden and the United States and elsewhere) are familiar to the audience, who engage emotionally with those they feel they know intimately enough to love or hate. For Debord, the star system was about a 'shallow seeming life', the denial of individuality (1983 §§ 60–61), but just how well or badly these celebrities on television behave under trying circumstances becomes a matter of great interest – and some amusement – to those who follow the show. They act out exaggerated ways of being together, knowing that they are being subject to moral judgement by millions of viewers.

The content of television can provide imagery in people's minds in the same way as ordinary life can and, because the perceptual origins of television are the same, there is likely to be a common stock of mental images. Each viewer's intentional orientation to these shared images will be idiosyncratic to some extent, but there will be shared aspects of each persona, including its good and bad qualities, that will make up the affect and knowledge associated with the mental image. The imaginary is not a shared set of precise contents, even to the extent that a language – including its grammar and lexicon – is, but audiovisual media, especially television, make available a realm of images that in substance are shared even if what is held at the level of the imaginary is variable.

The concept of the imaginary that I want to take forward from Lacan and Sartre refers to images that are perceived by many people as spectators who then hold them in their minds as quasi-sensible and irreal images. The imaginary may be employed at a conscious level of awareness, as when a relative or celebrity is recognised and with it flows knowledge and affect; we know the person's name, whom they are connected to, what sorts of situations they are likely to be in and how we feel about them. But the process may be much less conscious, less willed, less to do with focused attention and much more fluid than Sartre's account suggests. Image consciousness operates in the course of routine and habitual activities as we recognise and compare, recall and engage with the real objects that we come face to face with. The account of typifications that comes from Alfred Schutz (1974 – see Chapter 5, pp. 112–6) refers to a stock of knowledge made up of typical objects in typical situations with typical ways of acting or behaving. The imaginary is an equivalent aspect of the shared, collective memory but as the repository of images rather than the names, descriptions and understandings of typifications. Schutz's typifications are dynamic, with objects and people having relevance and motivation and the moral imaginary works in a similarly dynamic way so that imagery has relevance to ways of acting and being, ways of choosing to act and ways of judging the actions of others. The imaginary contains possibilities and likelihoods; what has been perceived in the past, what usually happens and what would be a breach of what is expected. It is not the product of reason or of thought, and though it may contain words or text, it is not discursive and cannot be simply translated into another mode such as speech or writing.

The social imaginary

The conceptualisation of the imaginary from Sartre and Lacan focuses on the memory and imagination of the individual, but Cornelius Castoriadis and Charles Taylor have both developed rather different notions of a shared and collective 'social imaginary'. Castoriadis wanted to counteract the 'functional-economic' point of view in political analysis that treats institutions as simply fulfilling social functions (1987, p. 115). He develops Lacanian psychoanalysis to provide a more social and political account of the imaginary that derives from the human capacity 'to see in a thing what it is not, to see it other than it is' and is made up of images that represent something (Castoriadis, 1987, p. 127). The imaginary is not simply non-real because there are layers of ritual and symbolic meaning connected to the 'real-rational' realm of material and functional existence. Castoriadis's concept of the imaginary is of a complex of sedimentary layers rather than a uniform realm; the 'radical imaginary' is the capacity for engaging in the non-real and the 'actual imaginary' is what surrounds the functional core of social institutions and is further made up of the 'central imaginary', the 'secondary imaginary' and the 'peripheral imaginary'. He argues that Marx's analysis of political economy is too oriented to the real-rational of institutions and so overlooks the mechanisms by which new needs are created.

It is, Castoriadis suggests, through the realm of the imaginary that new needs are discovered which then need to be satisfied:

> Humanity has hungered and does even now hunger for food, but it has also hungered for clothing and then for something other than

last year's fashions; it has hungered for automobiles and television; it has hungered for power and for saintliness; it has hungered for asceticism and for debauchery; it has hungered for love and brother-hood but it has also hungered for its own cadavers, hungered for feasts and for tragedies, and now it appears to be hungering for the Moon and the planets. (Castoriadis, 1987, p. 145)

What Castoriadis lists as the hungers of human beings sounds like a menu for the content of television broadcasting. If my argument that television is one of the key repositories of the social imaginary is right, then it is a place where creativity and innovation are displayed as well as the confirmation of tradition through repetitions and remakes. It is in the realm of the social imaginary that symbolic forms can be modi-fied and altered to make them appropriate for the society as it is now. There is a demand for new creations on television, new situations and new characters, but these are invariably revisions of previous forms. Much of the content of television is simply repeated, replaying old stories against new settings, but the television schedules are also full of new versions of old stories in a variety of formats. 'New' does not have to mean 'original' and as in all fields of innovation involves modifying and combining things that have already been made (Arthur, 2009). As I write, 'new' dramatised versions of stories based on Agatha Christie's 'Miss Marple' detective, dealing with multiple cadavers, and Richmal Crompton's schoolboy character 'William Brown', looking for feasts and avoiding tragedies, have recently been shown.[2]

The social imaginary that exists beyond the individual's imagination is for Castoriadis a sphere in which a society provides the answers to fundamental questions such as 'Who are we as a collectivity? What are we for one another? Where and in what are we? What do we want: what do we desire: what are we lacking?' (1987, pp. 146–7). These are the sorts of questions that are traditionally answered within religions, but they are also the stuff of ordinary everyday life, and the symbolic answers can be of enormous importance to the people within the society. Social identity, the order of nature or the 'world-order', the order of society and in particular its system of classes and the way that needs are defined, is always cultural and varies between one society and another. The symbolic explorations of these topics change within one society over time and may be different in different parts of the same society. Castoriadis uses the example of food, a material that seems to be a 'naturally determined' system of needs, values and desires but in fact varies so that what is a source of disgust in one culture is a delicacy in another. The values and choices that shape and specify needs and the significance of scarcity are shaped through symbolic means in the shared imaginary of the culture – this is, of course, a topic that is regularly featured in television programmes in which 'celebrity chefs' encourage us to combine foods in new ways and think of the foods of other cultures differently.

What Castoriadis includes within the 'social imaginary' is what I am calling a 'moral imaginary', a symbolic network of ideas through which a society makes sense of itself, in terms of its material and practical relations and in terms of how relations between people in the society should be conducted. This is not an ethical system from which rules of behaviour could be derived, but rather a set of possible connections between behaviours and values. Castoriadis points to the traditional role of religion in sustaining the social imaginary and its use of symbols to provide answers to difficult questions that are irreducible to rationality or to function. A crisis in modernity arises because of the tendency to presume that the logic of institutions can be explained by reference to functions and rational modes of action, while the imaginary and its underlying symbolic elements are overlooked or suppressed.

Religions have a much-reduced role in relation to the actual imaginary in modern, industrialised societies, but team sports such as the various versions of football (soccer, rugby, American football) have taken over some of their functions. In many modern countries, the passionate devotion of followers, the symbolism of colours and insignia and the collective effervescence of co-present singing, cheering and emotional discharge are quasi-religious. The showing of football matches on television contributes to sustaining a social imaginary because the spectacle of the event provides an emotional release and a sense of identity for many people who are not co-present. Its key characters – players, managers, trainers, referees and so on – provide exemplars of moral behaviour that are shared through the social imaginary. Certain actions – such as whether a ball was touched by hand or whether a collision between players was intended, careless or an accident – provide moments of great moral import both within the specific occasion and as demonstrations of standards of honesty and good behaviour. Castoriadis sees the social imaginary providing a 'world-order', the relative place of the objects within the world and particularly in relation to human beings. The symbolic order of things follows rational lines but 'arranges them and subordinates them to significations which themselves do not belong to the rational order ... but to the imaginary' (Castoriadis, 1987, p. 149).

The contribution of mediated symbolic exchange to the moral order and the social imaginary was famously articulated in Benedict Anderson's *Imagined Communities* (2006 [1983]). Anderson was trying to grasp the 'anomaly' of nationalism as a cultural artefact in which shared language and values bring people who are never likely to know or interact with each other, together into a shared sense of belonging as an imagined political community. The possibility of such a community was, he said, dependent on the technology of communication: '…innovations in the fields of shipbuilding, navigation, horology and cartography, mediated through print-capitalism, was making this type of imagining possible' (Anderson, 2006, p. 188). For Anderson, print capitalism made available media – books, pamphlets, newspapers and journals published by privately owned organisations and sold to recoup costs and make a profit – that enabled the exchange of ideas between people who had never met. He said that the newspaper and the novel 'provided the technical means for "re-presenting" the *kind* of community that is the nation' (Anderson, 2006, p. 25). The stories in a newspaper are connected, if by nothing else, through their juxtaposition on the page under a common title and a date. Different types of events in different places, involving different people are tied together in print as having occurred at the same time. The novel provides a narrative in which events are shown to be simultaneous in a way that enables connections between the actions of characters who never meet and at first glance appear to be independent. The newspaper achieves simultaneity through a 'calendrical coincidence' of peoples and events that are dispersed geographically and produce a sense of social connection.

Both the novel and the newspaper are commodities that are exchanged and circulated, but it is the newspaper that Anderson calls an extreme form of the book, a 'one-day best seller' sold 'on a colossal scale but of ephemeral popularity' (2006, pp. 34–5). The next-day obsolescence of the newspaper emphasises the specific temporal immediacy of 'news', that makes simultaneous the vicarious experience of disconnected people and events, to produce a 'mass ceremony' that is at once silent and private and at the same time replicated by thousands or millions of others. Television has carried on this role of connecting people who will never meet. It continues the role of constituting the regions of a nation-state as an imagined community by linked items in magazine programmes such as BBC's *One Show* (see Chapter 1). But it also links people across the world, and even across time, who will never share the same physical space but who find moral issues in common through audiovisual representation (see, for example, the discussion of 'distant suffering' in Chapter 7).

Television is only mentioned by Anderson in passing as an extension of newspapers; he does not seem to think of it as a different medium with a different mode of communication. Charles Taylor also has virtually nothing to say about the role of television except as an extension of the press, but he has recently used the concept of a 'social imaginary' to refer to 'the ways people imagine their social existence, how they fit together with others, how things go on between them and their fellows, the expectations that are normally met, and the deeper normative notions and images that underlie these expectations' (2004, p. 23). Taylor's version of the social imaginary has a moral/political flavour because it includes ideas about society as a whole. He argues that after the wars of religion, a new vision of moral order emerged in the 17th century based on the idea of human beings as rational, sociable and collaborating agents. Its political dimension was of natural rights that are built on mutual obligations following Locke's contractual theory of society 'as existing for the (mutual) benefit of individuals and the defense of their rights' (Taylor, 2004, p. 4).

For Taylor, there are three dimensions to the modern social imaginary – the public sphere, the economy and the practices of democracy – that displaced the centrality of religious thought. Through exchanging ideas in the public sphere, exchanging goods in the economy and exchanging sovereignty over each other in a political realm, people's moral relationships were transformed. They began to expect each other to demonstrate responsibility and mutual recognition, which led to challenges and disputes as well as agreement and harmony. In this new moral order, the interests of individuals began to be recognised in the social world of production and exchange as the 'purposes of a chain of beings enmesh[ed] with each other' (2004, p. 75). Science took on the role of revealing this new order of the social world as well as uncovering the order of the natural world. Taylor's social imaginary leads to political responses and a realm of rational debate in the public sphere in which there is a 'discourse of reason outside of power, which is nevertheless normative for power' (2004, p. 91). As I argued in Chapter 6, television provides a realm of discourse for the public sphere which can be more or less rational during news and current affairs programme but which can also have an emotional and 'private' orientation through chat shows and shows that explore personal issues raised by ordinary guests. These aspects of exchange in the public sphere also contribute to the social and moral imaginary.

Television as a social imaginary

It is strange that neither Anderson nor Taylor explores the role of tele-vision in the creation of imagined communities or social imaginaries (though the connection has been made by others, see, for example, Abercrombie and Longhurst, 1998, p. 115). In its various forms, the small screen has pushed the simultaneity of the newspaper and the novel aside as it draws on both the narrative form of the novel and the segmented and fragmented form of the newspaper in various types of programmes. There are news programmes that have a series of 'stories' with journalists who are often trained and experienced as print journal-ists presenting news and current events in a form that is directly related to that of the newspaper (for example, *The One Show*). And there are dramas, especially serial dramas, whose content and characters often originate in novels, that reveal the connections between the actions of different people at different times. The soap opera provides a half-way point, in which events follow the temporality of *now*; of news, of events that have just occurred, albeit the 'now' of a fictional and contained locality (for example, *EastEnders*). But soap opera also uses the narrative device of sequences of simultaneous, but apparently disconnected, sub-plots to show connections that would not be apparent from showing one continuous situation.

Not only does television replicate and modify the media effect of simultaneity that the novel and the newspaper achieved over the previous century and a half, it does so with the added realism of moving images and contemporaneous ambient sound that reinforce both the unfolding causality of action in time while also providing what Ellis (2002) calls the 'superabundance' of the image (see Chapter 5, p. 100). Photographic representation cannot avoid showing more extraneous detail than the programme makers ever intended, especially when drama and fiction are situated in existing real settings. Television on occasion achieves massive viewing figures for major social events, such as cup final matches, Royal weddings and funerals and celebrations such as those for the millennium. Simultaneously watched television may no longer have the same impact in the industrialised countries that it did in the second third of the 20th century – multiple channels and multiple platforms mean that content is not so determined by the moment of broadcast as by the consumer's choice. But still, the medi-ated content, whether broadcast, seen as a repeat or even as a boxed set some months later, is viewed by a vast audience as precisely the same content. Television, even better than the novel or the newspaper, can make what is distant feel co-present as its images are taken up in the social imaginary, the store of shared images that reflect various aspects of life.

Anderson's argument about the imagined community of nationality depends on the sharing of a language – he calls it 'print language' – that is common to the people of a given state. This means that dialects and even local languages have to become subordinated to a received language that is learnt and used by everyone. Insofar as this has happened, it has paved the way for electronic media but, of course, the residual imperial languages of Europe – especially English, French and Spanish – spill over the nation state boundaries. The distribution of broadcast media has tended to be within the nation-state, although programmes are sold from country to country, dubbed where necessary, sometimes adapted and remade. The ubiquity of the nation state and its impact on political processes that led to the development of what had been tribal collections or local ethnic groups, meant that in 1983 Anderson was able to claim, 'In a world in which the national state is the overwhelming norm…nations can now be imagined without linguistic commonality…' (2006, p. 135). He recognised that at least part of the reason for this is that the technology of radio and television has enabled multi-lingual broadcasting that could cheaply repeat a unifying message to different linguistic groups. Even more important is the pre-linguistic communication of the audiovisual moving images that draw people into the same event; a football match, a street demonstration or a war zone are all recognisable without text or speech, and not much in the way of speech or text is needed to understand what is going on.

As well as a moment of communication that connects the disparate viewers into some sort of community with a shared social imaginary, television has become its own archive of imagery. Television programmes will reprise earlier programmes, as we saw with both *The One Show* and *Crimewatch* in Chapter 1, and the use of archive film from a variety of sources is integrated into shows, like the use of footage of the death of President Kennedy on *Madmen* (AMC) and the many instances of both familiar and obscure footage in Adam Curtis's documentary series such as *All Watched Over by Machines of Loving Grace* (BBC Two). Jacques Derrida and Bernard Stiegler (2002) explore the issue of a national archive and the politics of memory that this entails; how can moving images be stored and made accessible and searchable? Since it is not possible to store everything, the decisions about what should be stored and so available for re-showing becomes political in a way that has yet to be fully explored. The television archive is itself a repository of moving imagery that can be called on to refresh those parts of the shared imaginary that are fading – but it depends on who has the right of 'inspection' of the archive of recordings (Derrida and Stiegler, 2002, p. 31).

The impartial spectator

Television distributes quasi-sensible and irreal images that can be contained in a social imaginary shared by many in a cultural community that crosses nation-state boundaries. I have argued that the perspective of the spectator of these images is more than the passive consumer whom Debord worried about, so what can those who draw on the social imaginary use it for? The television viewer is not simply a passive viewer of moving images but is put in the role of the moral spectator, one who is not involved and who has no direct interest in the outcome of the events they see. Luc Boltanski's (1993) concept of the moral spectator (see above, pp. 160–1) comes from Adam Smith's (1976 [1759]) moral philosophy. Smith's approach contrasts with much of the philosophical tradition that focuses on abstract principles of virtue and duty in that he began from the idea of 'sympathy', the interest that one person has in the situation of another:

> As we can have no immediate experience of what other men feel, we can form no idea of the manner in which they are affected, but by conceiving what we ourselves should feel in the like situation. (Smith, 1976, p. 60)

Smith develops the idealised and internalised notion of the 'impartial spectator' who is distinct from the uninvolved 'ordinary spectator' (Boltanski, 1993, p. 40). Smith's approach begins with intersubjectivity, with feelings and with what will produce a sense of propriety amongst a people. Rather than isolate qualities or behaviours that should be proscribed or commended, he builds his theory around the relations between people that constitute society. This perspective coincides with the phenomenology of television that I have derived from Husserl and Schutz through the concept of apperception in which understanding of the other person's experience depends on emotionally connecting it to one's own experience (see Chapter 5, pp. 112–3).

One of the most remarkable features of Adam Smith's *Theory of Moral Sentiments* (1976) is the role that he gives to the 'impartial spectator' within every human being. The impartial spectator is an aspect of the mind that draws on a disinterested perspective; it requires the individual to put themselves in the position of someone else, someone who knows what they know, including what they know about the culture they live in, but does not respond according to their particular interests. The ingeniousness of this idea is that it recognises that in order to be a member of society the individual has to be aware of what other people think and particularly, what they think about certain types of actions and behaviours. Specifically, people know what is right and what is wrong. It may feel 'instinctive' and indeed that is often how people will explain their moral judgements by offering not reasons but the justification that they know in their heart-of-hearts, in their bones, or that their instincts tell them, that such and such an act of someone is wrong. Something similar to this idea is contained in the symbolic interactionist idea of the 'generalised other', that component of the social self that George Mead (1934, p. 154) describes as being an internalised version, not of an individual being, but of the common response of other people. There is also a resonance with Charles Cooley's 'reflected or looking glass self', which is itself a sort of mature version of Lacan's mirror phase on which he bases his concept of the imaginary:

> As we see our face, figure, and dress in the glass, and are interested in them because they are ours, and pleased or otherwise with them according as they do or do not answer to what we should like them to be; so in imagination we perceive in another's mind some thought of our appearance, manners, aims, deeds, character, friends and so on, and are variously affected by it. (Cooley, 1964 [1922], p. 184)

In fact, Smith had used the same metaphor for a very similar purpose: 'This is the only looking-glass by which we can, in some measure, with the eyes of other people, scrutinize our own conduct' (1976, p. 112). There is even a connection with the internalised parental censure of Freud's (2010 [1923]) 'super-ego' through which the anticipated judgements of specific others become an internalised constraint on behaviour. Perhaps the most pertinent connection, however, is the connection between Smith's 'impartial spectator' and Durkheim's concept of the *'conscience collective'* (1933 [1893] – see Chapter 3, pp. 48–9). Both are ideas that connect the response of the individual's mind to their awareness and understanding of the minds of the others in the society around them, and both are ideas that refer to the shared moral response of a social group that is not based on reason or reflection. The French '*conscience*' carries connotations of consciousness and awareness, as well as conscience; a perceptual response as much as an internal one. '*Collective*' carries connotations of being shared by the social group, so the meaning of Durkheim's phrase is very close to what I mean by 'moral imaginary'. Durkheim's interest is in collective feeling, the solidarity of the society and what holds it together, whereas Smith was interested in the basis for morality within the person. My concept of 'moral imaginary' is somewhere between the two.

Smith's impartial spectator is internal to the individual, a version of their own conscience, but, as D. D. Raphael points out, the approval or disapproval of oneself:

> ...is an effect of judgements made by spectators. Each of us judges others as a spectator. Each of us finds spectators judging him. We then come to judge our own conduct by imagining whether an impartial spectator would approve or disapprove of it...Conscience is a social product of common social feeling.' (Raphael 2007, p. 35)

The idea of the spectator is important because behaviour is *viewed* as a spectator – this part of the self is not an active agent, takes no responsibility and indeed does not engage in a cognitive act of weighing things up or looking at pros and cons. The impartial spectator is an aspect of a self, viewing his or her own behaviour with the eyes of other people, as they would see it, and any judgement is the one that they would make without needing to reflect or reason. The impartial spectator takes up an imaginary position, one that cannot really exist, since it has the eyes of one person but the response of a collective or general other:

> We endeavour to examine our own conduct as we imagine any other fair and impartial spectator would examine it. If, upon placing ourselves in his situation, we thoroughly enter into all the passions and motives which influenced it, we approve of it, by sympathy with the approbation of this supposed equitable judge. If otherwise, we enter into his disapprobation, and condemn it. (Smith, 1976, p. 110)

The impartial spectator is a construct of imagination, not of cognition, reflection, reason or thought. It is a construct that emerges in the person as they become socialised, not through some creative act by them, nor as part of some practice or set of actions oriented towards their interests. The impartial spectator is the expression of the moral imaginary within the person as they draw upon the moral sensibilities that are taken to be normal in the society in which they live.

Smith's is a remarkably sociological view of the resources that the individual draws upon in making judgements of all sorts. Beauty, character, conduct and the basis for passions, joys and sorrows are all learnt from the others around the person. And just as one person judges others, so that person recognises that others will, and do, judge him – and this leads him to internalise and anticipate that external judgement. There is something of a separation within the person of the impartial spectator and the agent behind specific actions that Smith recognises – the judge is not precisely the same person as the judged. It is, of course, quite possible that a person can ignore the judgement of the impartial spectator and act in accordance with their desires and interests. There may be a rationalisation that says that 'no one else will know' or that 'the ends justify the means' – if free will exists, then every individual must have the capacity to ignore not only the wishes of others but also their own relevant internal judgements. Being motivated by the possibility of receiving praise or of being blamed, is not, of course, to act in accord with what one knows to be right – it is a mode of self-interest that might just conveniently fit with the judgement of others.

Raphael points out that the 'man within', as Smith sometimes refers to the impartial spectator, is not responding with actual social attitudes, rather with what the person takes the social attitudes to be. This is how the person can misjudge the way others would see things, under- or over-estimating the significance of an action or its consequences to others. And the reaction of others may not be the same; there will always be some variation in response and judgement, different spectators arriving at different conclusions. So the impartial spectator within the person is drawing on how *they* would feel if they were an impartial spectator, how would *they* respond to a particular course of action if it were not theirs but that of someone else?

There is an implicit idea of personal character in Smith's understanding of the moral sentiments because they are not present at birth and have to be learnt. The individual must develop their own impartial spectator through making judgements of others' behaviour and then using them to reflect on their own. The separation of person as agent and person as impartial spectator means that emotions or interests can determine action unless sufficient strength of character has been developed to apply constraints in line with the judgement of the impartial spectator. The child cannot be expected to have as well-developed a sense of the judgement as an adult because it is a matter of gaining

(2009, p. 56). It would be wrong to claim that the majority of contents of television should be considered as artworks – but some can be, and others might aspire to be. There is no reason why programme makers should not seek to use the medium to show programmes that have the potential for creating the 'vibrations' and the sense of 'aesthetic community' that Rancière writes about. Television may most usually be seen (especially when politicians from the political parties approve of its content) as a medium that fosters consensus, but it could be used to show possibilities that are not necessarily approved of by risking what Rancière calls 'dissensus…a conflict between two regimes of sense, two sensory worlds' (2009, p. 58). There are auteurs – for example, on UK television Paul Abbot, Alan Bleasdale, Adam Curtis, David Hare, Jimmy McGovern, Steven Poliakoff and Dennis Potter – who have attempted to create a sort of 'dissensual figure' by using the medium of television in unusual ways to raise issues about modern life that are complex and not normally discussed. Allowing provocative and aesthetically challenging programme makers access to broadcast and distribution is important for maintaining the breadth of imagery that feeds into the moral imaginary.

Conclusions

If the imaginary is for the individual a storehouse of imagery that connects the person with the world, it is from the mirror stage onwards social, populated as it is with people who are personally important as well as people and things that have been encountered. Some of the imagery that constitutes the imaginary may, in Sartre's terms, be irreal and only have a tenuous link with reality; television may be a source of both real and irreal imagery for the imaginative memories of individuals. The idea of a social imaginary, a shared sense of how the world is and what is important in it, is developed by Castoriadis and Taylor from Lacan's psychoanalytic concept. The social imaginary coheres the society, giving it a sense of collective identity, and is important for political and cultural institutions to be able to work. What I have called the 'moral imaginary' is an aspect of the social imaginary; those images that are relevant to the moral life of people within the society. For Castoriadis, the social imaginary includes the values and concepts that enable an economic and institutional life of the society to be carried on, and for Taylor it is the political and legal dimensions of the social imaginary that are most important. While the substance of the imagery may often be the same, my emphasis on the moral imaginary is focused on the possibilities for behaviour and interaction that are conducive to a shared social life. The moral imaginary contains the various values and attitudes, lines of action and styles of speech and gesture that individuals may consider, as well as a sense of the consequences and the social sanctions that may ensue.

Television is just one of the media through which the moral imaginary of contemporary societies is sustained. Stories, plays, novels, and writing in magazines, newspapers and blogs are literary resources, while painting, illustrations, photographs and films provide visual media that contribute to the moral imaginary. Aural media, including songs, poetry, speech and music can be shared 'live' or through radio, discs, the Internet and other means, and the so-called 'social media' blend text, image and music as does television. What makes television distinctive is the way that it combines visual and aural media that are often derived from literary resources. (Dennis Potter's 1986 series *The Singing Detective* was a high point in the use of television to bring many of these different channels of cultural mediation together at once.) All of these different modes of communication, whether local or 'mass', 'live' or recorded, create a 'moral culture', but television's combination of sound with moving visual images and the sheer variety and volume of content that it distributes, means it has become a key resource in contributing to the moral imaginary of modern societies.

Despite competition from other means of communication – particularly the reciprocal mediation of the Internet and the mobile phone – television continues to provide a model of shared culture that the other means of communication draw on. The contribution of television programmes to the repository of imagery that reflects and refreshes the moral order of a society, is reinforced by the repetition of programmes, the replaying of movies, the dramatisation of novels and the regular commentary on other cultural forms. The traces left in the minds and imaginations of individuals of the instances of 'good' and 'bad' are as likely to have arrived there through having watched television as through any other means. As I have argued in previous chapters, the moral content of television is not determinative of the moral beliefs or judgements of the people in society. The content of television is consumed as a flow, but it also accumulates as a pool or repository through which viewers can create their own flow. Unlike the hierarchically-controlled communication of religions, the medium of television, unless it is no more than an organ of a state propaganda, allows for a variety of different moral perspectives. These perspectives can interact with each other, stimulating discussion and debate that does not lead to a single set of ethical principles. Instead, the moral imaginary provides a resource from which people can find support and justification for different ways of acting. But the moral imaginary does not determine how individuals should act; they must interpret, judge and choose their own way. This is why it is more useful to think of television as contributing to a 'moral imaginary' than to think of it sustaining a 'moral order'.

9
What's Good on Television?

How frustrating! We have got so far, and still I haven't told you what the morality is that you can see on the small screen in your home. I've written about all sorts of social theories and philosophical ideas, reminded you of some programmes you might have seen and mentioned some you have heard of but never watched. But I haven't spelt out what the messages are, what the sum of morality on our screens is. So, is it good or is it bad? Which are the good programmes, the good channels? Which are the immoral depictions on the small screen? Which channel, programme, director, writer is producing the best morality for the audience? What are the criteria for judging good from bad television … in terms of morality, that is?

These are reasonable questions that are worth asking, and if you are asking them, then this book won't have been useless. But there are no answers that I can give; it is up to the viewer to engage consciously with what is on the small screen by thinking about the morality that is being promoted or undermined. It is the intelligence of the viewer that will filter the imagery and its meanings as they contribute to the impartial spectator within us all. What I have been arguing is that the variety of audiovisual moving images on the small screen offers a wide range of different moral possibilities; they are not, in the main, simple moral tales that tell you how to act or to be. This means that the medium itself cannot be 'good' or 'bad', but it also means that no single programme is good or bad in moral terms because what viewers take away from their viewing always depends on what they bring to it. Part of the pleasure in watching is to be faced with the conundrum of how to act and to be, in the light of the different patterns of action and consequences that are shown.

Any behaviour that has consequences for other people is of moral significance, which is why it is so difficult to specify the limits of the moral imaginary. Actions that lead to the death of someone else are of major importance, which is why death is such a recurrent theme on television programmes. Often what is at issue is not the approbation of an act of killing but the confusion over who is responsible, what mitigating circumstances there are and what other human costs are consequent on dealing with the killer. Sexual, procreative and family relationships are probably next in behavioural importance as morally significant acts, and they are often intertwined with each other and death. The mores surrounding appropriate sexual relationships and the degree to which they can be acknowledged in public changed over the 20th century, and television, both as actuality and in drama, has been involved in charting and responding to changes in what is acceptable. The more complex moral obligations to family members (parents to young children, children to their ageing parents) are another common theme in which expectations have changed over the course of a couple of generations.

Fundamental religious and philosophical beliefs are less of an issue for many people in western cultures and have become less significant in mainstream television (though the strength of feeling within various religious groups has led to the development of TV channels devoted to particular beliefs). On the other hand, monetary relationships and in particular the capacity for using money to buy goods, property and a lifestyle became an increasingly important cultural theme during the 20th century for the mass media. Politics, like economics, constitutes a set of social relationships that are irreducible to morality, but both entail moral values that are at various times more or less important. The obligation of those in power to abide by the rules that they expect others to follow, for example, led to the MPs expenses scandal in the UK in 2009. Perhaps the newest and most confusing theme of morality is that of the relationship of humans to the environment. It is confusing because it is a relationship between people that is mediated by the material stuff of the world, and the chains of connection are long and slow; the action of one person at one time may affect the life possibilities of someone else in a different place or time, but it is as yet difficult to be sure what the consequences are. It is a theme that periodically appears on television in a science show explaining climate change, a polemic on the need to reduce consumption or an account of the loss of species diversity.[1] References to the environment, particularly in economic and political contexts, are increasing, especially around 'ethical consumption', and it is emerging as a theme in the moral order with laws and prescribed practices around recycling and pollution.

Death, sex, religion, money, politics and the environment are some of the general themes that we will find in the moral imaginary, but it is not necessarily structured thematically. For example, 'care' is a moral issue that is entailed in all of these topics but may also constitute a topic or theme on its own. The amount and way that care is shown for others is fundamental to all morality, and yet it is difficult to grasp in the abstract beyond the context of particular relationships. (This is precisely the reason that 'distant suffering' is such a knotty problem – as we saw in Chapter 7.) In a similar way, professional relationships constitute, as Durkheim (1992) recognised, a particular set of moral issues that cut across all the moral themes I have mentioned.

Since it is not a moral or ethical system, the moral imaginary contains not only the current norms or standards of best practice, it also contains all the possible transgressions and sanctions. The variety of ways of breaching norms will each be associated with values indicating their degree of moral transgression. The values may be broadly shared throughout the culture that shares the moral imaginary, but they will always be inflected by personal and social characteristics (for example, of class, race, gender or age). What the moral imaginary will contain is the range of possible lines of behaviour, and it will provide a suffi-cient sense of shared values such that the 'impartial spectator' within everyone can draw on them to make judgements about one's own and others' behaviour.

I have argued that television cannot but present its audience with morality; by showing behaviour, both actual and fictional, and showing the consequences of that behaviour, the small screen gener-ates dynamic imagery that contributes to the moral imaginary of contemporary society. That society is no longer a nation state, and the moral imaginary is shared across many parts of the world. Socialisation through the small screen is continuous and can keep pace with a society whose membership, boundaries and structures are rapidly changing along with its moral order. Other media, including literature, comics, magazines and novels, contribute to the moral imaginary, but television makes a particular contribution because it provides complex audiovisual moving images in which the consequences of situations and actions are *shown* as they play out. Unlike literary forms, television is a direct and full representation of reality that does not need to be decoded to make sense, and it does not require sophisticated understanding or education to contribute to the moral imaginary.

Television has developed technologically, and its capacity to show the world to viewers through the small screen has been enhanced by 'televisual' techniques. Though the techniques risk, through aestheticisation, increasing the distance between the viewer and what is shown, televisuality has enabled the medium to be more ambiguous in the way it shows moral possibilities. The medium of television is itself moral in that there are ethical standards to be maintained, and there can be direct consequences to showing distant suffering as viewers are caught up in a televisual witnessing of the world. The realm of the imaginary is something that is human, and we all participate in as individuals; how we share it helps to shape our society and make it what it is. The fundamental idea of Adam Smith's 'impartial spectator' presumes that we can extend from our own experiences through the realm of the imaginary to make judgements about our own and others' behaviour. The realm of the imaginary is not exclusively moral, although its contents are almost always of more or less relevance to guiding our future actions and responding to those of other people. Television provides a continual stock of dynamic imagery that is shared through the practices of viewing so that it is both part of individuals' imagination and a shared imaginary.

The more that the small screen provokes and challenges its viewers without putting them off, the more useful the moral imaginary will be for sustaining a flexible and moral order. It is possible to imagine that the ideas and possibilities shown in programmes would become too diverse and contradictory for it to continue to attract a mass audience – the audience would simply turn off. This is why radical ideas or extreme versions of moral possibilities find it difficult to get on television and why programme makers are wary of being seen to be too much on one side rather than another.

A few programmes do challenge moral issues in a quite direct way as, for example, *Terry Pratchett: Choosing to Die* (BBC Two) did. This was a very personal documentary programme broadcast in June 2011 that explored what is involved in deciding when to end one's life. The author, who was suffering from Alzheimer's Disease at the time, wanted to find out how he might seek help if he wanted to choose to die and avoid the lack of dignity that the late stages of the disease usually involve. He talked to a number of people who were faced with a similar dilemma and visited the Dignitas Clinic in Switzerland, where he was present – as were the cameras – during an assisted death. It is difficult to imagine the programme having been made ten years earlier because of its challenge to the religious ethic of the sanctity of life with its duty to strive to sustain one's own and others' lives at all costs. The programme provoked debate on television and in the press and no doubt in a number of homes and workplaces; the debate would have extended beyond those who sat and watched the programme. The ideas explored were not new, but they were made more present and alive through the visible presence of the people who took part. The programme reached no 'conclusion' about whether assisted dying was a good or bad thing (and Pratchett did not make any decision for himself on camera). Apparently by coincidence, assisted dying was also a narrative theme in the ITV soap opera *Emmerdale* in June 2011, when Aaron Livesey helped his lover, Jackson Walsh, who was paralysed from the neck down, to die by drinking a cocktail of drugs. The fictional version exaggerated the visible emotion as against the actuality version, but both challenged the dominant legal and religious perspectives on assisted dying.

Television can be good at showing debate and discussion ... up to a point. There was a *Newsnight* special discussion following *Terry Pratchett: Choosing to Die* that allowed a number of different and dissenting views to be put forward and debated. Nonetheless, the programme and the BBC were accused of an 'orchestrated campaign' by some who argued that the broadcaster had an obligation to provide balanced reporting. The documentary, especially when put together with the *Newsnight* discussion, does however demonstrate the medium's capacity to engage in moral reasoning without arriving at an ethical judgement. The audiovisual medium means that discussion is not abstracted from the emotional, situational and personal aspects that contribute to reasons that the audience can then take into account. Such programmes clearly contribute to the 'public sphere' even if their viewing figures are low and there are limits to the number of such 'serious' programmes that can be included in the schedules.

If the issue of assisted dying has led to a few programmes (a BBC news report accompanied a woman to the Dignitas Clinic in 2006 and an earlier programme on Sky in 2008 had also shown the death of a man at the Dignitas Clinic), it remains the case that most television programmes confirm the norms of the legal and moral order. The possibilities and consequences of different lines of action that are explored tend to be within a range that is tolerated within the culture at the time – as the moral order shifts, so will the range of what is explored on television. Programme makers control what television programmes are made, and they make decisions about how to invest their time and money that are always concerned with sustaining an audience. The complex teamwork involved in getting a programme made and distributed means there are many checks and balances that will limit extremes; it is far easier for a single author to explore radical issues through a literary text whether on paper or on the Internet. Although we should not expect television programmes to be at the very forefront of cultural change, scouting out new possibilities, they may contribute to the vanguard once ideas have been expressed, and they will certainly contribute to the main discussion in the public sphere of ideas and alternative realities.

I have suggested that the sheer volume of television is part of its strength as a major contributor to the moral imaginary; the great variety of programmes that flow from broadcasters promises a range of different perspectives on different moral themes. But there is a risk that the plurality of ideas and perspectives becomes narrowed precisely because there are multiple channels and platforms. In its early days, 'television' was one thing, one channel whose flow was what the family watched and what was talked about at work and in social settings the next day. The increasing tendency towards 'niche' channels that cater to particular tastes and ideas means that while the medium may be diverse, what viewers watch may not be. In the early days of UK television, the single channel was too narrow, and a social elite controlled its content, but it did set aesthetic and intellectual standards that have largely endured into the multi-channel environment.

The range of channels and the easy availability of films and television series on DVD together with the recordings stored on DVRs and the Internet mean that viewers can choose nowadays what to watch and when, and with multiple screens, multiple choices can easily be made in the same household. While viewer choice is an important indicator of what television is acceptable and appreciated, it would be disappointing if it became the main determinant of what programming was commissioned or re-commissioned. Choosing what to watch, rather than simply joining the broadcast flow, means viewers are less likely to be exposed to situations, experiences and opinions that they are not already familiar with. It also means that they are less likely to confront those perspectives with which they disagree and which would provoke reflection on why they disagree. On the other hand, the technology that has led to the relative ease with which television content can be stored and recovered is creating an interesting archival problem as well as maximising viewer choice. The dynamic imagery that has been channelled through the small screen can literally become a repository as well as a moral imaginary if it is stored in an accessible way. If the archive is complete and searchable, and if all viewers have the 'right of inspection' (Derrida and Stiegler, 2002, p. 31), it can be used to replenish the moral imaginary through re-viewings and re-interpretations. The archive would, like a copyright library, provide an historical resource that could be used to understand the evolving nature of the moral imaginary.

The breadth of choice available to viewers can work against variety and difference if it leads to what Eli Pariser (2011) calls the 'filter bubble', a media space created around the individual by robotic filters that control what is allowed in to be watched, read and listened to. He is writing about the Internet rather than television, but he is concerned that 'personalisation' of the media will lead to the destruction of the public sphere as the filter technology of computer systems increasingly means that 'Our media is a perfect reflection of our interests and desires' (Pariser 2011, p. 12). One of the worrying capacities of filter technology is that it is invisible; we do not see the algorithms working to select what comes up on our online searches. Currently, the television viewer usually chooses what to watch from a published schedule, but TiVo and other systems can be set to record television output that fits in with what has been previously recorded, and a website like Amazon.com can offer a selection of DVDs according to previous purchases. As the range of choice expands, so the consumer gratefully accepts the help of systems that seem to offer the most attractive options. What Pariser warns us of is the 'advertar', or robotic assistant with a human persona that knows 'more about you, more precisely, than your best friend' and hails you in cyberspace to sell you something...or perhaps to tell you what you should watch on television (2011, p. 194). The filter bubble threatens to lead to an antisocial world that isolates the individual with their choices, linking them only to those people who share a particular interest. It would protect everyone from all those things and people they don't already know about – all the unknown unknowns that make life interesting.

There are still some very high-volume viewing events such as *The X Factor* (ITV1), *Strictly Come Dancing* (BBC One) and some of the soap operas that achieve between five and ten million viewers in the UK, but it is becoming more difficult to achieve a mass audience for a programme at a single time. The loss of the mass audience puts in question the sociality of television and its potential to create an imagined community; even as a one-to-many mode of communication, it has served well in bringing people together around events and issues (for example, *Cathy Come Home* – BBC, 1966, 9/11, royal weddings and funerals, New Year and millennium celebrations and the World Cup). As different people increasingly watch different things, or even watch the same thing at different times, the sociality that the medium affords is undermined. The sense of a shared experience, albeit a mediated one, that regular programming such as *Z cars* (BBC, 1962–1978) and *'Till Death Us Do Part* (BBC, 1965–1975) achieved in an earlier era is at risk of being lost.

It is important that a variety of material continue to find its way onto the small screen – this is the key role of commissioning editors and producers in encouraging new topics and styles rather than simply going for a formula that has worked in the past. But it is also important that sufficient viewers be engaged for the medium to contribute to debate in the public sphere. The representation of different experiences and opinions and maintaining a variety of views and perspectives on what is appropriate and acceptable behaviour should be treated as criteria for television content by broadcasters and programme makers. I have been frustratingly unspecific about the morality that is carried on television, but this is because I have wanted to emphasise the importance of television as a medium to the form of late-modern society. Television is a good thing for society because it communicates dynamic imagery that can help to sustain a shared moral imaginary across a large number of people, supporting a shared sense of an accepted moral order and sustaining what I have called a 'mediated solidarity' (see Chapter 3, pp. 50–1).

I have put forward a positive view of the social role of the small screen seldom shared by academic commentators who are keener on criticising (or sometimes praising) the content of particular programmes. But there are serious threats to the important social role of television. One is the risk of interference by political authority into the content of programmes and who makes them. Politicians will always have interests in how they and their policies are represented, and these cannot be detached from the broader interest of allowing the media to be free to criticise and represent a range of views. Even the expectation of 'balance' can lead to restrictions in the freedom of programme makers to express ideas and opinions, and it can also lead to the presentation of views that have little support or basis. A further threat is from those pressure groups acting on behalf of particular sectors of society who would seek to impose censorship, perhaps for religious reasons or, even more arbitrarily, on 'moral' grounds. Another risk is the abandonment of television to 'market forces' that would shape content according to commercial interests. The greatest threat here is when the concentration of capital with ever-larger multi-national corporations leads to a unitary policy and avoids radical or innovative ideas. The state and the market will always affect what is broadcast, but neither should determine all content. If too extensive, these two sources of intervention in the making and distributing of television would lead to a diminution of the breadth of imagery available to the moral imaginary. The moral imaginary needs to be continually replenished with new images and new possibilities, and for television to continue to play its part in shaping late-modern societies, it must be free of any dominant form of moralising.

Notes

1 Introduction – The Small Screen and Morality

1. In the UK, 93 per cent of homes have a digital television set, and the average number of hours of television watched per day was as high as ever in 2010, at 4 hours, up by 23 minutes per day since 2005 (Ofcom 2011a). In the US, the comparable figure was just over 5 hours per day in 2009 (Nielsen 2009). In both settings, traditional television outstrips the attention given to all other electronic media combined.
2. I will expand on what I mean by the 'phenomenology' of television in Chapter 5, but for now I am using the term to refer to the way television is experienced by human consciousness.
3. Smell-o-vision is periodically reinvented but has yet to achieve any success (for example, http://www.dailymail.co.uk/sciencetech/article-1322394/TV-Smell-o-vision-Turn-aroma-Nigella-Lawson-please.html, accessed 23 May 2011). 'Feely' television is perhaps less far away as haptic robotic devices and computer consoles like the Wii bring touch and movement into the process of interacting with a small screen, alongside sound and vision.
4. In 2008, the *New York Times* reported, 'Despite sagging home sales, rising unemployment and record-high gas prices, the number of TVs shipped to retailers in the United States and Canada jumped 26 per cent compared with the first quarter of this year, and 28 percent year over year, to a total of 9.3 million units. (That's the number of all types of TVs shipped; consumer sales generally lag about a month behind shipments.)', http://bits.blogs.nytimes.com/2008/08/15/despite-economy-tv-sales-continue-to-rise/, accessed 11 August 2011.
5. BARB Top Thirty Viewing Figures, June 2008, available at: http://www.barb.co.uk/, accessed 26 September 2011.
6. The latest edition was published in February 2011, available at: http://stakeholders.ofcom.org.uk/broadcasting/broadcast-codes/broadcast-code/, accessed 26 September 2011.It includes sections to protect under-18 viewers with a 'watershed' at 9pm, protect all viewers from 'harm and offence', maintain 'accuracy and fairness', protect privacy and sets out rules surrounding elections and referendums.
7. '"Documentary" is the loose and often highly contested label given, internationally, to certain kinds of film and television (and sometimes radio programmes) which reflect and report on the "the real" through the use of recorded images and sounds of actuality' (Corner 1996: 2).
8. PCSOs wear similar uniforms to the police but have a different training and recruitment process with a limited range of powers and duties. They were introduced to supplement the work of local police authorities in the UK in 2002.

9. I will distinguish between 'actuality', which is taken to be a representation of reality, and fiction or drama, which is presented as 'made up', even if based on real-life experiences. The French word *actualité* usually refers to what we would call news or current affairs – that which has actually happened – and I will use 'actuality' in the same way to refer to television that purports to represent what has actually happened (for example, news, sport, current affairs, magazine programmes and documentaries). I also note that Scannell (1996: 111) discusses authenticity and the emergence of 'actuality broadcasting' that enters everyday lives outside the studio. The presentation of 'reality' is often 'made up' in the sense that it is an interpretation of reality, and there is no guarantee of truth (see Chapter 7).

10. BARB Top Thirty Viewing Figures, June 2008, available at: http://www.barb.co.uk/, accessed 26 September 2011.

2 Morality on Television

1. Here and in all later quotations, the emphasis is in the original unless otherwise stated.

2. I am drawing on a Martha Nussbaum's (2009) list of Aristotelian virtues but avoiding some of the tricky ones, such as 'greatness of soul', 'expansive hospitality' and 'friendliness'. The Lone Ranger always seemed to me a rather single-minded cold fish and a busybody who wouldn't have earned much respect from Aristotle.

3. The term 'reality' should, of course, be treated with great caution because it is usually applied to programmes in which a cast of people, who are not trained and who have no script, are brought together to improvise a scenario specified by rules and requirements in a setting controlled by the programme's producers. This televisual construction of 'reality' is not, of course, the same as reality – events and activities that would have happened anyway – that is then televised.

4. 'Cosa Nostra' (literally 'our thing') is another name for the Mafia, apparently used by its members. See http://news.bbc.co.uk/1/hi/world/europe/7086716.stm, accessed 11 September 2009.

5. 'Diegesis' here refers to the fictional world in which the characters, situations and events occur.

3 Sociology and the Moral Order

1. http://www.youtube.com/watch?v=HECMVdl-9SQ, date accessed 11 September 2011.

2. http://stakeholders.ofcom.org.uk/broadcasting/broadcast-codes/broadcast-code/, accessed 8 August 2011.

3. It is important to note that the 'Other' that Bauman invokes derives from Levinas and is rather different from the 'Other' derived from Lacan that Žižek invokes and we will meet in Chapter Seven.

4 Televisuality: Style and the Small Screen

1. Barthes (1968: 76–78), of course, argued that there was no such thing as colourless, neutral, writing degree zero, and the same argument could be made about zero-degree television; mimesis always adds and takes away, style is never really absent.
2. Feedback using electronic means, including email, blogging, and especially Twitter, has recently put programme makers in much more direct and rapid contact with their audience; the content of a live programme can even be shaped by such responses.
3. Vivian Sobchak's (1992: 15–16) phenomenological theory of film discusses the three metaphors for the screen as a window, a mirror and a picture frame.
4. 'I inscribe a quadrangle of right angles, as large as I wish, which is considered to be an open window through which I see what I want to paint.' (Alberti 1956: 56).
5. Putting both window and screen at the end of a rectangular tube would make both images effectively monocular. If the television image were to use autostereoscopic 3D technology, which is a technical possibility although not yet widely available, the rectangular tunnel could be dispensed with.
6. At the current response and refresh rates of LCD screens, the complex fluidity of animal movement is most likely to leave a noticeable 'ghost' trace.
7. For example, BBC television's 'Walking with Dinosaurs', broadcast in 1999, used computer-generated graphics and animatronics overlaid onto video of real settings to imitate those documentaries that show a level of detail of the minutiae of animal life that would be impossible for a human body to observe directly.
8. The possibility of a superhuman being able to absorb the content of multiple television screens showing different channels was one of many provocative ideas in Nicholas Roeg's 1976 film 'The Man Who Fell to Earth'.
9. 'What is being offered is not, in older terms, a programme of discrete units with particular insertions, but a planned flow, in which the true series is not the published sequence of programme items but this sequence transformed by the inclusion of another kind of sequence, so that these sequences together form the real flow, the real "broadcasting"' (Williams 1992: 84).
10. In one of those interesting reversals of media, some cinemas in the UK have begun to 'broadcast' live dramatic performances from the National Theatre and the Royal Opera House.

5 The Phenomenology of Television

1. Walter Benjamin notes the traditional presumption that the masses seek *distraction* whereas art commands the *concentration* of the spectator – but for him 'reception in a state of distraction…is symptomatic of profound changes in apperception' brought about by the mechanical reproduction of moving images that herald a break with the cult value of auratic art and open the way for a distracted mass to be absorbed by art (1973: 233).

2. Television had only just been invented when he was writing in 1928, but he did mention it as a 'new approach' to 'the public mind' (Bernays 2005: 167).

3. "The outfits of some of the dancers were revealing, with limited coverage of the buttocks, and were of a sexualised nature because they were based on lingerie such as basques, stockings and suspenders," the regulator added. "The routine ... had a number of simultaneous, sexualised elements concentrated into a relatively short period of time and there was therefore a cumulative effect." Quotations from the Ofcom report in the Guardian, 20 April 2011, available at: http://www.guardian.co.uk/media/2011/apr/20/x-factor-cleared-rihanna-christina-aguilera-dance-routines, accessed 22 April 2011.

4. I will return to the notion of 'witness' in Chapter 7.

6 Society and the Small Screen

1. 'Sociation is the form (realized in innumerably different ways) in which individuals grow together into a unity and within which their interests are realized. And it is on the basis of their interests ... that individuals form such unities' (Simmel 1971: 24).

2. BARB (Broadcaster's Audience Research Board), available at: http://www.barb.co.uk/facts/tvOwnershipPrivate, accessed 7 June 2011.

3. BARB 'Trends in Television Viewing' 2010, published February 2011.

4. The recent fictional drama series *Borgen* (Danish Broadcasting Corporation, 2010–) has explored the role of television news in political life much more closely in these ways than earlier political dramas, such as *The West Wing* (NBC 1999–2006) or *The State Within* (BBC 2006).

7 Mediating Morality

1. http://www.mediawatchuk.org.uk/, date accessed 15the July 2011.

2. http://www.vlv.org.uk/pages/about.php, date accessed 15th July 2011.

3. A seven minute version of the original report with Buerk's powerful commentary is available on the BBC website at: http://news.bbc.co.uk/onthisday/hi/correspondents/newsid_2626000/2626349.stm, date accessed 18 July 2011.

4. In Chapter 2 there is a discussion of the related concept of moral relativism.

5. What Fiske is describing is the traditional mode of television without self-conscious style that Caldwell calls 'zero-degree television' (1995: 15 – see Chapter 4, p. 72, above).

6. Although there is an affinity between Žižek's concept of 'big Other' and my 'moral imaginary', the repository of the former is much larger than an imaginary that is specifically concerned with morality, mores, norms, attitudes and behaviours.

7. There is an uncomfortable resonance between those who enter the Big Brother house and those who enter a refugee camp; both submit to a form of cultural 'bare life' in which they are unable to act politically or socially but simply become characters on a screen whose bodies are available for viewing, rather than killing, without sacrifice, abandoned to a distant, unaccountable sovereign power (Agamben 1995: 83).

8. Quoted in a review by Martin Buttle on the website of 'impactt' an organisation dedicated to ethical trade and improving working conditions, available at: http://www.impacttlimited.com/2008/05/14/blood-sweat-and-t-shirts-coming-face-to-face-with-child-labour/, date accessed 26 July 2011.

8   Television and the Imaginary

1. Writing in 1940, television was not culturally significant, but Sartre, who wrote not only plays but also cinema screenplays, makes no mention of the role the moving image in contributing to the imaginary.
2. The Agatha Christie story has been reset from its original time period of the 1920s to the 1950s, and the characters reinvented or renamed – Miss Marple was not even in the original version of the story. Crompton's William Brown featured in a series of books, the first of which was published in 1921, and the character and stories have been the focus of four previous television series, three feature films and a number of radio series and readings from the books.

9   Conclusions

1. For a digest of examples, see: http://www.thedailygreen.com/living-green/blogs/celebrities/best-green-television-shows-461008, date accessed, 21st September 2011. And for a themed Internet TV channel, see http://www.green.tv/?set_location=en, accessed 21 September 2011.

References

Abercrombie, Nicholas and Longhurst, Brian (1998) *Audiences: A sociological theory of performance and imagination* (London: Sage).

Adorno, Theodor and Horkheimer, Max (1979 [1944]) 'The Culture Industry: Enlightenment as Mass Deception', in *Dialectic of Enlightenment* (London: Verso).

Adorno, Theodor (1991 [1954]) 'How to look at television', in T. Adorno, *The Culture Industry* (London: Routledge).

Agamben, Georgio (1998) *Homer Sacer: Sovereign Power and the Bare Life* (Stanford California: Stanford University Press).

Alberti, Leon Battista (1956 [1435]) *On Painting*, New Haven (Connecticut: Yale University Press).

Alasuutari, Pertti (1996) 'Television as a Moral Issue' in P. L. Crawford and S. B. Hafsteinsson (eds.) *The Construction of the Viewer: Media Ethnography and the Anthropology of Audiences* (Højbjerg, Denmark: Intervention Press).

Anderson, Benedict (2006 [1983]) *Imagined Communities: Reflections on the Origin and Spread of Nationalism,* (London: Verso).

Ang, Ien (1985) *Watching Dallas: Soap Opera and the Melodramatic Imagination* (London: Methuen).

Archard, David (1998) 'Privacy, the public interest and the prurient public', in M. Kieran (ed.), *Media Ethics* (London: Routledge).

Aristotle (1996) *Poetics*, trans. Malcolm Heath, London: Penguin Books

Aristotle (1999) *Nicomachean Ethics,* Translated by Terence Irwin, Second Edition, Indianapolis: Hackett Publishing Inc.

Arthur, W. Brian (2009) *The Nature of Technology* (London: Penguin Books).

Baier, Kurt (1995) *The Rational Order and the Moral Order: The Social Roots of Reason and Morality* (Chicago: Open Court).

Bandura, Albert (1968) 'What TV Violence Can Do to Your Child' in O. N. Larson (ed.) *Violence and the Mass Media* (New York: Harper & Row).

Bandura, Albert (1978) 'A social learning theory of aggression', *Journal of Communication*, Vol. 28(3), 12–29.

Bandura, Albert (1994) 'Social Cognitive Theory of Mass Communication' in J. Bryant and D. Zillman (eds.) *Media Effects: Advances in Theory and Research* (Hillsdale, New Jersey: Lawrence Erlbaum Associates).

Barr, Rachel (2010) 'Transfer of learning between 2D and 3D sources during infancy: Informing theory and practice', *Developmental Review*, Vol. 30, 128–154.

Barthes, Roland (1968 [1953]) *Writing Degree Zero* (New York: Hill and Wang).

Barthes, Roland (1977 [1961]) 'The Photographic Message' in *Image-Music-Text* (Glasgow: Fontana).

Barthes, Roland (1993 [1980]) *Camera Lucida* (London: Vintage).

Baudrillard, Jean (1981) For a Critique of the Political Economy of the Sign (New York: Telos Press).

Baudrillard, Jean (1991) *The Gulf War did not take place* (Sydney, NSW: Power Publications).

Baudrillard, Jean (1983) *Simulations* (New York: Semiotext(e)).

Baudrillard, Jean (1994 [1981]) *Simulations and Simulacra* (Ann Arbor: The University of Michigan Press).

Baudrillard, Jean (2002a) *Screened Out* (London: Verso).

Baudrillard, Jean (2002b) *The Spirit of Terrorism* (London: Verso).

Bauman, Zygmunt (1989) *Modernity and the Holocaust* (Cambridge: Polity).

Bauman, Zygmunt (1992) *Intimations of Postmodernity* (London: Routledge).

Bauman, Zygumnt (1993) *Postmodern Ethics* (Oxford: Blackwell).

Bauman, Zygmunt (1995) Life in Fragments: Essays in Postmodern Morality (Oxford: Blackwell).

Bauman, Zygmunt (2002) *Society under Siege* (Cambridge: Polity).

Benjamin, Walter (1973) 'The Work of Art in the Age of Mechanical Reproduction', in *Illuminations*, (London: Fontana Press).

Bentham, Jeremy (1987 [1824]) 'An Introduction to the Principles of Morals and Legislation' in J. S Mill and J. Bentham, *Utilitarianism and other essays* (London: Penguin Books).

Berger, Peter L. and Luckmann, Thomas (1966) *The Social Construction of Reality*, New York: Anchor Books.

Bernays, Edward (2005 [1928]) *Propaganda* (New York: Ig Publishers).

Bignell, Jonathan (1997) *Media Semiotics: An Introduction* (Manchester: Manchester University Press).

Bird, S. Elizabeth (2003) *The Audience in Everyday Life: Living in a Media World* (London: Routledge).

Bogdan, Robert (1988) *Freak Show: Presenting Human Oddities for Amusement and Profit* (Chicago: University of Chicago Press).

Boltanski, Luc (1993) *Distant Suffering: Morality, Media and Politics* (Cambridge: Cambridge University Press).

Bourdieu, Pierre (1984 [1979]) *Distinction: A Social Critique of the Judgement of Taste* (London: Routledge).

Butler, Jeremy G. (2010) *Television Style* (London: Routledge).

Caldwell, John Thornton (1995) *Televisuality: Style, Crisis and Authority in American Television* (New Brunswick, New Jersey: Rutgers University Press).

Caldwell, John Thornton (2008) Production Culture: Industrial Reflexivity and Critical Practice in Film and Television (Durham: Duke University Press).

Calhoun, Craig (ed.) (1996) *Habermas and the Public Sphere* (Cambridge, Mass.: MIT Press).

Carroll, Noël (1998) 'Is the medium a (moral) message?' in M. Kieran (ed.), *Media Ethics* (London: Routledge).

Carroll, Noël (2003) *Engaging the Moving Image* (New Haven: Yale University Press).

Castoriadis, Cornelius (1987 [1975]) *The Imaginary Institution of Society* (Cambridge, Mass.: MIT Press).

Chouliaraki, Lilie (2006) *The Spectatorship of Suffering* (London: Sage).

Chouliaraki, Lilie (2011) '"Improper distance": Towards a critical account of solidarity as irony', *International Journal of Cultural Studies*, Vol. 14(4), 363–381.

Cooley, Charles Horton (1964 [1922]) *Human Nature and the Social Order* (New Jersey: Schocken Books).

Corner, John (1995) *Television Form and Public Address* (London: Hodder Headline).

Corner, John (1996) *The Art of Record: A Critical Introduction to Documentary* (Manchester: Manchester University Press).

Corner, John (1999) *Critical Ideas in Television Studies* (Oxford: Clarendon Press).

Cottle, Simon (2006) 'Mediatized rituals: beyond manufacturing consent' (*Media, Culture and Society*, Vol. 28(3), 411–432.

Couldry, Nick (1996) 'Media discourse and the naturalization of categories' in R. Wodak and V. Koller (eds.) (2008) *Handbook of Communication and the Public Sphere* (Berlin: Mouton de Gruyter.

Critcher, Chas (2003) *Moral Panics and the Media* (Buckingham: Open University Press).

Crossley, Nick and Roberts, John M. (2004) *After Habermas: New Perspectives on the Public Sphere* (Oxford: Blackwell).

Cumberbatch, Guy et. al. (1987) *The portrayal of violence on British television* (London: BBC Data Publications).

Dahlgren, Peter (1995) *Television and the Public Sphere: Citizenship, Democracy and the Media* (London: Sage).

Dant, Tim (1991) *Knowledge, Ideology and Discourse: A Sociological Perspective*, London: Routledge.

Dant, Tim (2003) *Critical Social Theory: Culture, Society and Critique*, London: Sage.

Dant, Tim (1999) *Material Culture in the Social World: Values, Activities, Lifestyles*, Buckingham: Open University Press.

Dant, Tim (2005) *Materiality and Society* (Maidenhead, Berkshire: Open University Press).

Dayan, Daniel and Katz, Elihu (1992) *Media Events: The Live Broadcasting of History* (Cambridge, Mass.: Harvard University Press).

De Waal, Frans (2006) 'Morally Evolved: Primate Social Instincts, Human Morality and the Rise and Fall of "Veneer Theory"', in S. Macedo and J. Ober (eds.) *Primates and Philosophers* (Princeton, New Jersey: Princeton University Press).

Deans, Jason (2004) 'Ofcom hits out at Big Brother violence', *Guardian Media*, 18 October 2004.

Debord, Guy (1983 [1967]) *Society of the Spectacle* (Detroit: Black and Red).

DCMS (2006) 'Broadcasting: Copy of the Royal Charter for the continuance of the British Broadcasting Corporation', *Department for Culture, Media and Sport*, Cmnd 6925, available at: http://www.bbc.co.uk/bbctrust/assets/files/pdf/about/how_we_govern/charter.pdf, date accessed 12 August 2010.

Derrida, Jacques and Stiegler, Bernard (2002) *Echographies of Television* (Cambridge: Polity Press).

Dewey, John (2002 [1922]) *Human Nature and Conduct* (New York: Prometheus Books).

Dewey, John (1960 [1908]) *Theory of the Moral Life* (New York: Holt, Rinehart and Winston).

Dienst, Richard (1994) *Still Life in Real Time* (Durham, NC: Duke University Press).

Durham-Peters, John (2011) 'Witnessing' in P. Frosh and A. Pinchevski (eds.) *Media Witnessing: Testimony in the Age of Mass Communication* (London: Palgrave).

Durkheim, Emile (1933 [1893]) *The Division of Labour in Society* (New York: The Free Press).

Durkheim, Emile (1938 [1895]) *The Rules of Sociological Method* (New York: The Free Press).

Durkheim, Emile (1974 [1924]) 'The Determination of Moral Facts' in *Sociology and Philosophy* (New York: The Free Press).

Durkheim, Emile (1992 [1957]) *Professional Ethics and Civil Morals* (London: Routledge).

Durkheim, Emile (2002 [1925]) *Moral Education* (New York: Dover Publications).

Durkheim, Emile (2001 [1912]) *The Elementary Forms of the Religious Life* (Oxford: Oxford University Press).

Economist (2010) 'A special report on television', *The Economist*, 29 April 2010).

Ellis, Anthony (1998) 'Censorship and the media', in M. Kieran (wd.), *Media Ethics* (London: Routledge).

Ellis, John (1982) *Visible Fictions: Cinema, Television, Video,* London: Routledge.

Ellis, John (2002) *Seeing Things: Television in the Age of Uncertainty* (London: I.B. Taurus).

Ellis, John (2009) 'The Performance on Television of Sincerely Felt Emotion', *The Annals of the American Academy of Political and Social Science*, 625, 103–15.

Ellis, John (2011) 'Mundane Witness' in P. Frosh and A. Pinchevski, (eds.), *Media Witnessing: Testimony in the Age of Mass Communication* (London: Palgrave).

Feldman, Fred (1992) 'Kantian Ethics' in J. B. Steera (ed.) *Ethics: The Big questions* (Oxford: Wiley-Blackwell).

Felson, Richard B. (1996) 'Mass Media Effects on Violent Behavior', *Annual Review of Sociology*, Vol. 22, pp. 103–128.

Ferguson, Christopher J. (2011) Video Games and Youth Violence: A Prospective Analysis in Adolescents', *Journal of Youth Adolescence*, Vol. 40, 377–91.

Fiske, John (1987) *Television Culture* (London: Routledge).

Fiske, John and Hartley, John (1978) *Reading Television* (London: Methuen).

Fraser, Nancy (1996) 'Rethinking the Public Sphere: A Contribution to the Critique of Actually Existing Democracy', in C. Calhoun (ed.) *Habermas and the Public Sphere* (Cambridge, Mass.: MIT Press).

Freud, Sigmund (2010 [1923]) *The Ego and the Id* (Radford, Virginia: Wilder publications Ltd).

Frosh, Paul and Pinchevski, Amit (eds.) (2011) *Media Witnessing: Testimony in the Age of Mass Communication* (London: Palgrave).

Frosh, Paul (2011) 'Telling Presences: Witnessing, Mass Media and the Imagined Lives of Strangers', in P. Frosh and A. Pinchevski, (eds.), *Media Witnessing: Testimony in the Age of Mass Communication* (London: Palgrave).

Gaita, Raimond (1991) *Good and Evil: An Absolute Conception* (Abingdon, Oxford: Routledge).

Garnham, Nicholas (1996) 'The Media and the Public Sphere' in C. Calhoun (ed.) (1996) *Habermas and the public sphere* (Cambridge, Mass.: MIT Press).

Gauntlett, David (2004) *Moving Experiences: Media Effects and Beyond* (Second Edition) (Eastleigh: John Libbey).

Gerbner, George and Gross, Larry (1976) 'Living with Television: The Violence Profile', *Journal of Communication*, Vol. 26 (2), 173–199.

Gerbner, George; Gross, Larry; Morgan, Michael; Signorelli, Nancy (1994) 'The Cultivation Perspective', in J. Bryant and D. Zillman (eds.) *Media Effects:*

Advances in Theory and Research (Hillsdale, New Jersey: Lawrence Erlbaum Associates).

Gergen, Kenneth (2002) 'Technology, Self and the Moral Project' in J. E. Davis (ed.), *Identity and Social change* (New Brunswick, NJ: Transaction Books).

Gillespie, Marie (ed.) (2005) *Media: Audiences* (Maidenhead: McGraw Hill/Open University Press).

Ginsberg, Morris (1956) *On The Diversity of Morals* (London: Heinemann).

Goffman, Erving (1956) 'The Nature of Deference and Demeanor', *American Anthropologist*, Vol. 58 (3), 473–502.

Goffman, Erving (1961) *Asylums: Essays on the Social Situation of Mental Patients and Other Inmates* (Chicago: Aldine Publishing Company).

Goffman, Erving (1972) *Interaction Ritual* (Harmondsworth: Penguin).

Gouldner, Alvin W. and Peterson, Richard A. (1962) *Technology and the Moral Order* (Indianapolis, New York: Bobs Merrill).

Graham, Gordon (1998) 'Sex and violence in fact and fiction', in M. Kieran (ed.), *Media Ethics* (London: Routledge).

Gunter, Barry (1994) 'The Question of Media Violence' in J. Bryant and D. Zillman (eds.) *Media Effects: Advances in Theory and Research*, Hillsdale, New Jersey: Lawrence Erlbaum Associates).

Habermas, Jurgen (1990 [1983]) *Moral Consciousness and Communicative Action* (Cambridge: Polity Press).

Habermas, Jurgen (1992 [1962]) *The Structural Transformation of the Pubic Sphere* (Cambridge Polity).

Hawkins, Gay (2001) 'The ethics of television', *International Journal of Cultural Studies*, Vol. 4(4): 412–426.

Hill, Thomas E. Jr. and Zweig, Arnulf (2002) 'Editors' Introduction' to I. Kant, *Groundwork for the Metaphysics of Morals* (Oxford: Oxford University Press).

Hodge, Robert and Tripp, David (1986) *Children and Television: A Semiotic Approach* (Oxford: Polity).

Holmes, Su (2008) '"The viewers have...taken over the airwaves'? participation, reality TV and the approaching the audience-in-the-text', *Screen*, Vol. 49 (1), 13–31.

Horton, Donald R. and Wohl, Richard (1956) 'Mass communication and Para-Social Interaction: Observations on Intimacy at a Distance', *Psychiatry*, Vol. 19, 215–29.

Housley, William and Fitzgerald, Richard (2009) 'Membership categorization, culture and norms in action', *Discourse & Society*, Vol. 20 (3), 345–62.

Husserl, Edmund (1999 [1950/1933]) *Cartesian Meditations: An introduction to phenomenology* (Dordrecht: Kluwer Academic Publishers).

Iedema, Rick (2001) 'Analysing film and television: A social semiotic account of *Hospital: and Unhealthy Business*' in C. Jewitt and T. van Leeuwen (eds.), *Handbook of Visual Analysis* (London: Sage).

Ignatieff, Michael (1985) 'Is Nothing Sacred? The Ethics of Television' *Daedalus*, Vol. 114 (4), 57–78.

Ignatieff, Michael (1998) *The Warrior's Honor: Ethnic War and the Modern Conscience* (London: Chatto and Windus).

Inthorn, Sanna and Boyce, Tammy (2010) 'It's disgusting how much salt you eat!: Television discourses of obesity, health and morality', *International Journal of Cultural Studies*, Vol. 13 (1), 83–100.

James, William (1950 [1890]) *Principles of Psychology: Volume 2* (New York: Dover).

James, William (1927 [1899]) *Talks to Teachers* (London: Longmans, Green and Co).

Kackman, Michael; Binfield, Marnie; Payne, Matthew Thomas; Perlman, Allison and Sebok, Bryan (2011) *Flow TV: Television in the Age of Media Convergence* (London: Routledge).

Kant, Immanuel (2002) *Groundwork for the Metaphysics of Morals*, translated and by T. E. Hill Jr. and A. Zweig (Oxford: Oxford University Press).

Katz, Elihu and Lazarsfeld, Paul (1960) Personal Influence: The Part Played by People in the Flow of Mass Communications (New York: Free Press).

Kellner, Douglas (1981) 'Network Television and American Society: Introduction to Critical Theory of Television', *Theory and Society*, Vol. 10 (1): 31–62.

Kellner, Douglas (1990) *Television and the Crisis of Democracy* (Boulder, Colorado: Westview Press).

Kellner, Douglas (1995) *Media Culture: Cultural Studies, Identity and Politics Between the Modern and the Postmodern*, London: Routledge.

Kieran, Matthew (ed.) (1998a) *Media Ethics* (London: Routledge).

Kieran, Matthew (1998b) 'Objectivity, impartiality and good journalism', in M. Kieran (ed.), *Media Ethics* (London: Routledge).

Kirjnen, Tonny and Costera Meijer, Irene (2005) 'The moral imagination in primetime television', *International Journal of Cultural Studies*, Volume 8 (3), 353–74.

Kittler, Friedrich A. (1999) *Gramophone, Film Typewriter*, Stanford, California: Stanford University Press.

Klapper, Joseph T. (1960) *The Effects of Mass Communication* (New York: Free Press).

Korsgaard, Christine (2006) 'Morality and the Distinctiveness of Human Action', in S. Macedo and J. Ober (eds.) *Primates and Philosophers* (Princeton, New Jersey: Princeton University Press).

Krahé, B.; Möller, I.; Huesmann, L. R.; Kirwil, L.; Felber, J.; Berger, A. (2011) 'Desensitization to Media Violence: Links with Habitual Media Violence Exposure, Aggressive Cognitions and Aggressive Behavior', *Journal of Personality and Social Psychology*, Vol. 100 (4), 630–46.

Lacan, Jacques (1977 [1966]) *Écrits: A Selection*, translated by Alan Sheridan (London: Tavistock).

Lacan, Jacques (1979 [1977]) *The Four Fundamental Concepts of Psycho-Analysis*, translated by Alan Sheridan (Harmondsworth, Middlesex: Penguin Books.

Langer, Suzanne K. (1957) *Philosophy in a New Key: A Study in the Symbolism of Reason, Rite, and Art*, Cambridge, Massachusetts: Harvard University Press.

Lefebvre, Henri (1991 [1947/1958]) *Critique of Everyday Life, Volume I* (London: Verso).

Lefebvre, Henri (2002 [1961]) *Critique of Everyday Life, Volume III* (London: Verso).

Lidz, Victor (1984) 'Television and Moral Order in a Secular Age', in W. Rowland and B. Watkins (eds.) *Interpreting Television: Current Research Perspectives*, Beverley Hills: Sage.

Luckmann, Thomas (2007) 'Moral Communication in Modern Societies', *Human Studies*, Vol. 25(1), 19–32.

Lukes, Steven (2010) 'The Social Construction of Morality?' in S. Hitlin and S. Vaisey (eds.) *Handbook of the Sociology of Morality* (New York: Springer).

Lukes, Steven (2008) *Moral Relativism* (London: Profile Books.

Lury, Karen (2005) *Interpreting Television* (London: Hodder Arnold).

MacIntyre, Alastair (1985) *After Virtue (Second Edition)* (London: Duckworth).

McEnery, Tony (2006) *Swearing in English: Bad Language, Purity and Power from 1586 to the Present* (London: Routledge).

McLuhan, Marshall (1994 [1964]) *Understanding Media: The Extensions of Man* (London: MIT Press).

Mander, Jerry (1978) *Four Arguments for the Elimination of Television* (New York: Morrow Quill Paperbacks).

Manovich, Lev (2001) *The Language of the New Media* (Cambridge, Massachusetts: MIT Press.

Mead, George (1934) Mind, Self and Society: From the Standpoint of a Social Behaviourist (Chicago: Chicago University Press).

Merleau-Ponty, Maurice (1962) *Phenomenology of Perception* (London: Routledge).

Metz, Christian (1974 [1971]) *Film Language: A Semiotics of the Cinema* (Chicago: University of Chicago Press.

Meyrowitz, Joshua (1985) *No Sense of Place: The Impact of Electronic Media on Social Behavior* (New York: Oxford University Press).

Mill, John Stuart (1987 [1863]) 'Utilitarianism' in J. S. Mill and J. Bentham, *Utilitarianism and other essays* (London: Penguin Books).

Miller, Toby (2010) *Television Studies: The Basics*, London: Routledge.

Morley, David (1980) *The Nationwide Audience* (London: BFI).

Morley, David (1992) *Television, Audiences and Cultural Studies* (London: Routledge).

Nelson, Jenny L. (1986) 'Television and Its Audiences as Dimensions of Being: Critical Theory and Phenomenology', *Human Studies*, Vol. 9(1): 55–69.

Newcomb, Horace M. and Hirsch, Paul M. (1984) 'Television as a Cultural Forum', in W. Rowland and B. Watkins (eds.) *Interpreting Television: Current Research Perspectives* (Beverley Hills: Sage).

Nielsen (2009) *Nielsen A2/M2, 3 Screen Report: Television, Internet and Mobile Phone Usage in the US, 1st Quarter 2009*, Nielsen Company, available at: http://kr.en. nielsen.com/site/documents/A2M23ScreensFINAL1Q09.pdf, date accessed 26 September 2011.

Nielsen (2011) *The Cross-Platform Report*, available at: http://www.nielsen.com/ us/en/insights/reports-downloads/2011/cross-platform-report-q1–2011.html, date accessed 26 September 2011.

Nussbaum, Martha (1992) 'Non-relative virtues: An Aristotelian Approach' in J. B. Steera (ed.) *Ethics: The Big questions* (Oxford: Wiley-Blackwell).

Nussbaum, Martha (2009) 'Non-relative Virtues: An Aristotelian Approach' in J. B. Steera (ed.) *Ethics: The Big Questions*, (Oxford: Wiley-Blackwell).

Ofcom (2011a) *Communications Market Report:* UK, 04/08/11, available at: http:// stakeholders.ofcom.org.uk/market-data-research/market-data/communica-tions-market-reports/cmr11/, date accessed 26 September 2011.

Ofcom (2011b) *Broadcast Bulletin Issue number 179* 04/04/11, available at: http:// stakeholders.ofcom.org.uk/enforcement/broadcast-bulletins/obb179/, date accessed 26 September 2011.

Orwell, George (2004) *Nineteen Eighty-Four* (London: Penguin Modern Classics).

Pariser, Eli (2011) The Filter Bubble: What the Internet is Hiding from You (London: Penguin).

Patterson, Philip and Wilkins, Lee (2005) *Media Ethics: Issues and Cases (Fifth Edition)* (New York: McGraw Hill).

Plunkett, John (2004) 'Police investigate Big Brother fight', *Media Guardian*, Thursday 17 June, 2004.

Pojman, Louis P. (ed.) (2000) *The Moral Life: An Introductory Reader in Ethics and Literature* (New York: Oxford University Press).

Postman, Neil (1985) *Amusing Ourselves to Death* (London: Methuen).

Raney, Arthur A. and Bryant, Jennings (2002) 'Moral Judgement and Crime Drama: An Integrated Theory of Enjoyment', *Journal of Communication*, Vol. 52, 402–15.

Rancière, Jacques (2009) *The Emancipated Spectator*, London: Verso).

Raphael, D. D. (2007) *The Impartial Spectator* (Oxford: Oxford University Press).

Rawls, Anne Warfield (2010) 'Social Order as Moral Order' in S. Hitlin and S. Vaisey (eds.) *Handbook of the Sociology of Morality* (New York: Springer).

Richardson, Kay (2008) 'Specific debate formats of mass media' in R. Wodak and V. Koller (eds.) (2008) *Handbook of Communication and the Public Sphere* (Berlin: Mouton de Gruyter).

Saenz, Michael (1992) 'Television Viewing as a Cultural Practice', *Journal of Communication Inquiry*, Vol. 16(20): 37–51.

Sartre, Jean-Paul (2004 [1940]) The Imaginary: A Phenomenological Psychology of the Imagination (London: Routledge).

Scannell, Paddy (1996) *Television, Radio and Modern Life* (Oxford: Blackwell).

Schutz, Alfred and Luckmann, Thomas (1974) *Structures of the Life-World: Volume 1* (London: Heinemann Educational Books).

Schutz, Alfred (1971) 'Symbol, Reality and Society' in *Collected Papers, Vol. 1: The Problem of Social Reality* (The Hague: Martinjus Nijhoff).

Schulzke, Marcus (2010) 'Defending the morality of violent video games', *Ethics of Information Technology*, Vol. 12, 127–38).

Shilling, Chris and Mellor, Philip A. (1998) 'Durkheim, morality and modernity: collective effervescence, *homo duplex* and the sources of moral action', *British Journal of Sociology*, Vol. 49 (2), 193–209.

Silverstone, Roger (1994) *Television and Everyday Life* (London: Routledge).

Silverstone, Roger (2007) *Media and Morality* (Cambridge: Polity).

Skeggs, Bev (2005) 'The Making of Class and Gender through Visualizing Moral Subject Formation', *Sociology*, Vol. 39(5), 965–82.

Skeggs, Bev; Thumim, Nancy and Wood, Helen (2008) '"Oh goodness, I am watching reality TV": How methods make class in audience research', *European Journal of Cultural Studies*, Vol. 11(5), 5–24.

Smith, Adam (1976 [1759; 6th Edition 1790]) *The Theory of Moral Sentiments*, Edited by D. D. Raphael and A. L. Macfie (Oxford: The Clarendon Press).

Smith, Anthony (1985) 'The Influence of Television', *Daedalus*, Vol. 114 (4), 1–15.

Smith, B. R. (no date) 'The Lone Ranger', *Museum of Broadcast Communications*, available at: http://www.museum.tv/eotvsection.php?entrycode=loneranger, date accessed on 23rd September 2011.

Sobchak, Vivian (1992) *Address of the Eye: A Phenomenology of Film Experience* (Princeton: Princeton University Press).

Stevenson, Nick (2003) *Cutural Citizenship: Cultural Questions* (Maidenhead, Berks.: Open University Press).

Sumner, William Graham (1906) Folkways: A Study of the Sociological Importance of Usages, Manners, Customs, Mores and Morals (Boston: Ginn and Company).

Taylor, Charles (2004) *Modern Social Imaginaries* (Durham and London: Duke University Press).

Talyor, Paul A. (2010) *Žižek and the Media* (Cambridge: Polity).

Tester, Keith (1997) *Moral Culture* (London: Sage).

Tester, Keith (1999) 'The Moral Consequentiality of Television', *European Journal of Social Theory*, 2(4), 469–83.

Tester, Keith (2001) *Compassion, Morality, and the Media* (Buckingham: Open University Press).

Tudor, Andrew (1999) *Decoding Culture: Theory and Method in Cultural Studies* (London: Sage).

Turowetz, Jason T. and Maynard, Douglas, W. (2010) 'Morality in the Social Interactional and Discursive World of Everyday Life', in S. Hitlin and S. Vaisey (eds.) *Handbook of the Sociology of Morality*, New York: Springer).

Tyler, Imogen (2008) '"Chav Mum Chav Scum"', *Feminist Media Studies*, 8 (1), 17–34.

Urry, John (2000) Sociology beyond Societies: Mobilities for the Twenty-First Century (London: Routledge).

Weber, Max (1992 [1905]) *Protestant Ethic and the Spirit of Capitalism* (London: Routledge).

Williams, Raymond (1992 [1974]) *Television: Technology and Cultural Form* (London: Wesleyan University Press).

Wodak, Ruth and Koller, Veronik (eds.) (2008) *Handbook of Communication and the Public Sphere* (Berlin: Mouton de Gruyter).

Wuthnow, Robert (1987) *Meaning and Moral Order: Explorations in Cultural Analysis* (Berkeley: University of California Press).

Zack, E.; Barr, R.; Gerhardstein, P; Dickerson, K.; Meltzoff, A. N. (2009) 'Infant imitation from television using novel touch screen technology', *British Journal of Developmental Psychology*, Vol. 27, 13–26.

Zillman, Dolf and Bryant, Jennings (1975) 'Viewer's Moral Sanction of Retribution in the Appreciation of Dramatic Presentations', *Journal of Experimental Psychology*, Vol. 11, 572–82.

Žižek, Slavoj (2001) The Fright of the Real: Kristof Kieslowski between Theory and Post-Theory (London: British Film Institute).

Žižek, Slavoj (2002) *Welcome to the Desert of the Real* (London: Verso).

Žižek, Slavoj (2008) 'The Lacanian Real: Television', *The Symptom*, No. 6, June 10th, p. 38, available at http://www.lacan.com/symptom/, date accessed 26 September 2011.

Index